NEW DIRECTIONS FOR CHILD AND ADOLESCENT DEVELOPMENT

Science for Society

Informing Policy and Practice Through Research in Developmental Psychology

Ann Higgins-D'Alessandro
Fordham University

Katherine R. B. Jankowski
Fordham University

EDITORS

Number 98, Winter 2002

JOSSEY-BASS
San Francisco

SCIENCE FOR SOCIETY: INFORMING POLICY AND PRACTICE THROUGH RESEARCH IN DEVELOPMENTAL PSYCHOLOGY
Ann Higgins-D'Alessandro, Katherine R. B. Jankowski (eds.)
New Directions for Child and Adolescent Development, no. 98
William Damon, Editor-in-Chief

Microfilm copies of issues and articles are available in 16mm and 35mm, as well as microfiche in 105mm, through University Microfilms Inc., 300 North Zeeb Road, Ann Arbor, Michigan 48106-1346.

ISSN 1520-3247 electronic ISSN 1534-8687

NEW DIRECTIONS FOR CHILD AND ADOLESCENT DEVELOPMENT is part of The Jossey-Bass Education Series and is published quarterly by Wiley Subscription Services, Inc., a Wiley company, at Jossey-Bass, 989 Market Street, San Francisco, California 94103-1741. Periodicals postage paid at San Francisco, California, and at additional mailing offices. Postmaster: Send address changes to New Directions for Child and Adolescent Development, Jossey-Bass, 989 Market Street, San Francisco, CA 94103-1741.

New Directions for Child and Adolescent Development is indexed in Biosciences Information Service, Current Index to Journals in Education (ERIC), Psychological Abstracts, and Sociological Abstracts.

SUBSCRIPTIONS cost $79.00 for individuals and $180.00 for institutions, agencies, and libraries.

EDITORIAL CORRESPONDENCE should be sent to the Editor-in-Chief, William Damon, Stanford Center on Adolescence, Cypress Building C, Stanford University, Stanford, CA 94305.

Cover photograph by Wernher Krutein/PHOTOVAULT © 1990.

Jossey-Bass Web address: www.josseybass.com

Printed in the United States of America on acid-free recycled paper containing at least 20 percent postconsumer waste.

CONTENTS

EDITORS' NOTES

It is only in the last twenty-five years that psychologists in any numbers have become involved in interpreting research findings for policymakers and organizations. It is even more recently that psychologists have begun to frame and conceptualize their research questions to directly address issues of social and public policy. Recognizing the importance of this endeavor marks the advent of applied developmental psychology and applied developmental science as fields and as specializations of graduate study and training.

Science for Society: Informing Policy and Practice Through Research in Developmental Psychology presents four leaders in the evolution and expansion of child and family developmental psychology into the social policy arena. This volume is an embodiment of the spirit and tenets of applied developmental psychology as portrayed through the life work of these pioneers in child and adolescent development: Ed Zigler, Ruby Takanishi, Aletha Huston, and Robert Selman. Although not all began their professional lives as developmental psychologists, all were vitally interested in the life experiences of children and families. They became child developmental researchers unusually talented in revealing the implications of their work for understanding children's lives in situ. Their genius, however, lay not only in their research but also and especially in their commitment to enhance children's lives. Thus their professional lives illustrate both the roots and the goals of these new fields of applied developmental psychology and applied developmental science.

Edward Zigler and Sally Styfco document Zigler's cumulative impact and influence in shaping public thought about children's development. This portrait of his professional life shows how he has used psychological research and his forceful presence to create the forums from which he tirelessly advocates for policies and programs (especially Head Start) to meet children's needs. He encourages developmentalists to tell policymakers what is known and acknowledge what is not known in order to help shape policy.

Ruby Takanishi explores how the complex perspectives of her own marginality have enabled her to see opportunities unseen by others and to make substantial contributions across a range of developmental issues. Most of all, the openness of Takanishi's perspective from the margin shows in her strong and effective leadership in the world of child advocacy.

Aletha Huston takes us back to her own childhood, describing her parents' influence on her political and scientific thinking. Her professional life illustrates how she has helped to frame social policies about children's television and to address the needs of children in poverty by asking good research questions, using the best methods available, and being ready when political trends and times offer opportunities for influencing policy.

Robert Selman has lived his life in two worlds, the worlds of research and clinical practice. His professional life is a clear example that when research is done in the context of programs designed to promote children's development, its yield is double: both better services for children and enhanced theoretical understanding of developmental issues.

We are grateful to these leaders for sharing their life experiences and wisdom of the field by writing chapters and by participating in an earlier conference on applied developmental psychology, "Influential Lives: Four Developmental Psychologists Tell Their Life Stories," which was held at Fordham University in February 2000. Several Fordham faculty members also contributed to this volume. Their chapters are examples of social policy research issues and projects.

Research with ethnic groups and cultures or subcultures other than European Americans often still uses measures developed for European Americans. Nancy Busch-Rossnagel describes the logic, methods, and benefits of creating measures sensitive to both the cultures and the communities of specific research participants.

Many clinical interventions for at-risk youth are not developmentally sensitive. Aaron Hogue provides a framework and practical examples for using rigorous implementation research to create programs that are more developmentally appropriate interventions for at-risk adolescents. Ann Higgins-D'Alessandro argues that promoting adolescent development means optimizing teacher development. She analyzes the job of teaching, identifying and illustrating four necessary work conditions for fostering teacher development at schools.

Lonnie Sherrod (previously at William T. Grant Foundation and now at Fordham University) argues that building substantial connections between research and policy rests on seven considerations we must take seriously when doing research. He encourages psychologists to design research programs that combine the criteria of policy analysis, such as cost-benefit analysis, with developmental issues, such as appropriateness and continuity of services across the life span.

All the authors of this volume remind us that our work is conducted within the historical, demographic, and political configurations of our time and that research alone never determines policy. The life experiences and research offered here provide insights and encouragement to all psychologists whose goal is providing robust evidence relevant to framing social policies and creating programs for America's children and families.

We especially want to acknowledge the Fordham developmental psychology graduates and students and others across the nation who are in courses and programs that emphasize the necessarily complex relationships between good research and good social policy. As pioneers in this programmatic effort, we hope that the professional lives illustrated in this book will guide as well as root their own professional goals to enhance the lives of children and families and better the social conditions of our shared life.

Ann Higgins-D'Alessandro
Katherine R. B. Jankowski
Editors

Acknowledgments

I wish to express my deep regard for and gratitude to Kathy Jankowski. Kathy has been my partner and colleague in every way—planning and executing the very successful conference that became the basis for this book, and then developing and bringing it to fruition. Kathy added these projects to her full life as a graduate student, researcher, wife, and mother and kept us smiling most of the time. It has been a pleasure to have her at Fordham University.

I also thank my colleagues Nancy Busch-Rossnagel, Aaron Hogue, and Lonnie Sherrod for participating in the conference and writing chapters for this book. And I want to recognize my colleagues in the Fordham University Applied Developmental Psychology graduate program, Celia Fisher, Kathleen Schiaffino, Daniel Mroczek, and Reesa Vaughter (retired). Our working together over the past twelve years to build a strong graduate program and contribute to the development of this new field has been, and continues to be, exciting. I would also like to thank my husband Thomas for his support in my endeavors.

A.H.D.'A.

I would like to thank Ann Higgins-D'Alessandro, my mentor and guide through graduate school. I have been lucky to be mentored by someone who has belief in ideals, tenacity, and genuineness in addition to being a caring and warm person. She is and continues to be a role model for me and others who are working toward becoming applied developmental psychologists. I also want to thank Jeff, my husband; Marianna, my daughter; and Robert and Rose Bennett, my parents, for their unfailing support.

K.R.B.J.

ANN HIGGINS-D'ALESSANDRO is associate professor and director of the Graduate Program of Developmental Psychology at Fordham University, Bronx, New York.

KATHERINE R. B. JANKOWSKI is a research associate in the Graduate School of Social Services and an applied developmental psychology doctoral student at Fordham University, Bronx, New York.

1

This chapter documents the influential impact on America's social policy of one developmental psychologist's knowledge, presence, and advocacy. It is a story of spirit, tenacity, and hope throughout the struggle to improve the quality of life for America's children and families.

A Life Lived at the Crossroads of Knowledge and Children's Policy

Edward Zigler, with Sally J. Styfco

In the early 1960s, I was an ambitious young professor doing what most respectable psychologists at the time were doing—basic research. I had two, some would say detached, fields of interest. One was abnormal psychology. Many of you may not realize that I did my doctoral training as a clinical psychologist. The other research area was mental retardation, in which I became interested as a graduate student—that was the topic of my dissertation. While some of my seniors thought I just couldn't make up my mind, I did not see these areas as all that disparate. From a developmental perspective, I was beginning to understand various psychiatric disorders as related to the patient's premorbid level of functioning, which I interpreted as being their developmental level. Among children with cultural-familial retardation, my work was indicating that their functioning was similar to that of nonretarded children who were chronologically younger but at the same level of cognitive development. These two lines of thought eventually grew up to become developmental psychopathology and the developmental approach to mental retardation. The philosophy was the same: People's functioning is linked to their developmental level, regardless of their mental health status or rate of cognitive growth.

The Early Head Start Years

When the call came from Washington to work on planning Head Start, I thought they had the wrong number. Robert Cooke, a noted pediatrician, was forming a committee of experts to design a school readiness program for young children who lived in poverty. Cooke had heard me lecture on why retarded children often perform below what would be expected at their cognitive level. I theorized that because they had failed so often, their self-confidence and motivation to try were diminished. Cooke thought that poor children also experienced an inordinate amount of failure and would do better in school if they believed they had a chance to succeed. Thus he thought I had something to contribute to his committee. My only experience with economically disadvantaged populations was that the children with cultural-familial retardation in my studies were generally from poor families. I myself had grown up poor. So with some trepidation—and against the advice of some of my mentors, who were devoted to basic science and against applied science—I agreed to join what became the Head Start Planning Committee.

The committee was convened under the mission of the War on Poverty, an ambitious federal effort to help the poor help themselves to share more fully in the nation's resources and democracy. President Lyndon Johnson and Sargent Shriver, who was Johnson's chief strategist of the program, both believed that education was the vehicle for escaping poverty. Shriver, who had experience in the education field, thought that poor children did not do as well in school as their wealthier peers because they were behind when they started. Thus he envisioned a national intervention to help them begin school on a more equal footing.

With the exception of a few small experimental projects, there was little experience at the time to suggest how to meet the needs of economically disadvantaged preschoolers. The fourteen committee members had backgrounds in medicine, mental health, social work, and early childhood education, and we all felt that our own disciplines should be part of the intervention. I was the youngest, and probably noisiest, person on the committee. I also came to the meetings with a different perspective from most of the others. While the majority worked in applied fields, I was trained in empirical methodology, hypothesis testing, and theory building. Thus I injected into our deliberations an often unwelcome dose of scientific skepticism and demands that our plans be knowledge-based or at least piloted and carefully studied.

But Head Start was not meant to be a pilot program. I thought I had won a round by convincing my fellow planners that there was absolutely no evidence that our ideas would work so we should test them out, do some research, and use the results to inform our next step. I felt a sense of accomplishment when the group counseled Shriver that a small trial program should be run and evaluated before implementing Head Start on a national scale. President Johnson, however, was a true Texan with Texas-size ideas. He demanded that Shriver fire a major volley in the antipoverty war by starting out in a grand manner. When Head Start opened in the summer of 1965, which was just a few months after we finished our planning, more than half a million children attended.

I was troubled by this development. How could we subject all these children and their families to this hasty, unproven, indefinite program? I came to terms with what we insiders dubbed "Project Rush Rush" for two reasons. One is that the planners agreed that what we were doing would not be harmful. The children would get immunized, have their teeth and eyes checked, and eat nutritious meals. Their families would receive some services and be invited to participate in parent education classes and in their children's early schooling. Some of this might do some good, but I really didn't have particularly high hopes. Head Start was only a single summer program; children attended for just six or eight weeks before they started elementary school in the fall. No one on the planning committee believed we could break the cycle of poverty in such a short period of time. About all we were sure of was that preschool education, health care, and good nutrition would be beneficial, and the other services would at least do no damage.

The other reason I was able to accept bypassing the demonstration phase is that there really was no choice. Julius Richmond, who was my longtime mentor and would become Head Start's first director, tried to ease my fears by pointing out that we had this fleeting window of opportunity in which to roll out Head Start. The public and political climates were very favorable. There was the will, the finances, and the friends in the right places—namely, Johnson and Shriver. The conflict in Vietnam was beginning to stir, but it had not yet drained our spirits and resources. If we didn't do Head Start right then, it would never happen.

It also occurred to me that when we were asked to plan a program, the result was expected to be an operational program. This was my first realization about the different worlds of policymakers and academic psychologists. Academics devise ways to test ideas using small groups, repeating the process as they refine both the ways and the ideas. Policymakers are action-oriented. They mount real programs to serve real people and then go on to the next item on the social action agenda. It took me years to get used to that. Over time I came to accept that there will be occasions when psychologists do not think they have sufficient knowledge about a social issue to justify a policy recommendation, but they still know more than policymakers do. Laws will be made with or without us, so we might as well contribute what we do have.

The one round I really did win on the planning committee at first seemed to be a gigantic flop. True to my empirical training, I insisted that Head Start had to be evaluated so we could have some way of knowing if the program was accomplishing anything and some clues as to what we might do better. The majority of my fellow planners felt we were going to give children health care and some pleasant experiences—all good things—so there really wasn't anything to evaluate. I found an ally in Julie Richmond, who advised the committee that research and evaluation be added to the planning document. Payback time came when he told me to go ahead and develop the measures we would use.

I worked with Edmund Gordon, who was director of Head Start's evaluation component. We only had a little over a week to decide on what to measure and how. So I rushed back to my research lab at Yale, assembled a team of graduate students, and worked nonstop to develop some tests of Head Start's various objectives. With the little time we had, our instruments were never formally validated, so the scores were not a bit useful. Their only value was in representing a commitment to evaluation that would take some time to mature. That is happening today, and I guess I should be proud of my stubbornness in making research part of Head Start.

The Office of Child Development

My next foray in uncharted territory came just a few years later, when the Nixon administration asked me to serve as the first director of the Office of Child Development, which is now the Administration on Children, Youth and Families. While I had doubts about leaving academia to become a bureaucrat, I was attracted by the administration's expressed interest in early childhood. There was also the possibility that I would be able to help shape the child care component of the Family Assistance Plan, which was the president's proposal to help families leave welfare. Finally, no one really understood what this new child development office was supposed to be. It would be up to me to set its course—and thus the course of our nation's response to meeting children's needs. This was more than an opportunity to me. It was an obligation.

I saw the office as having three main functions. One was to operate Head Start. The second was to coordinate child and family services at the national level. Third was to continue the Children's Bureau's time-honored traditions of identifying problems and advocating for children and families. The Children's Bureau was created in 1912 and was now part of OCD. In addition to being director of OCD, I was also appointed chief of the Children's Bureau, and I wanted to emulate the efforts of former chiefs. A task I took on later, in conjunction with President Nixon's welfare reform proposal, was conceptualizing and developing plans to implement a national system of child care.

I had mixed success in each of these functions. It took a few decades of work after I left Washington to see real progress made on some of the goals I set so long ago. Beginning with Head Start, I learned shortly after arriving in Washington that the administration was entertaining plans to phase it out. So the fact that it is alive and well thirty years later represents a victory of sorts, with the hero being Elliot Richardson, my immediate superior as secretary of Health, Education and Welfare. My contribution toward saving Head Start was to respond to some negative publicity over the Westinghouse findings that the gains children made in Head Start purportedly faded away in the first few years of elementary school. My action plan was two-pronged: First, I had to clarify for policymakers and laypeople what Head Start was supposed to accomplish. Second, I tried to raise program quality, which had been uneven since the program's hasty beginnings.

Head Start's goals were a source of confusion from the beginning. Our planning document did not contain a straightforward goal but listed seven objectives, ranging from improved physical health and abilities to a better "sense of dignity and self-worth" for the child and family. This made it hard to explain the point of the program, which in turn made it easy for observers to focus on the one part of one objective that everyone understood—cognitive skills. For example, when President Johnson read a preliminary report that children's IQs went up by as much as 10 points after spending one summer in Head Start, he found something he could talk about. He proclaimed the program a resounding success because it made the children smarter. This set the stage for studies like the Westinghouse report and the hundreds that followed to focus on intelligence and achievement, ignoring Head Start's efforts to enhance school readiness, child health, and family functioning. And it set Head Start up for failure, since it could not attain the impossible goal of raising IQ scores significantly and permanently.

When I ran Head Start, I saw the confusion the planners inadvertently created and pushed for a simpler goal statement. After several years of operation, Head Start officially received the goal of social competence. Yet this construct did not mean much to the public and, as it turned out, meant too many things to professionals. Over the next ten years teams of scientists conducted three major efforts to operationally define social competence and develop an assessment battery for use in Head Start evaluations. Despite this significant amount of work, no practical results were achieved. Head Start continued to be misunderstood as a cognitive enrichment program and disparaged because the enriched IQs it produced did not last forever.

School Readiness as a Goal. In 1990, President Bush and the state governors adopted national education goals for the year 2000. Goal One is that all children will enter school ready to learn. Policymakers began to embrace preschool programs as the means of attaining universal school readiness. They renewed their friendship with Head Start and poured funds into it for expansion. I quickly saw the value of using terminology most people could understand.

Although I had long insisted that Head Start's goal is social competence, for young children this is exactly the same as school readiness. To be competent, a child must be effective in dealing with the environment and be able to meet age-appropriate social expectancies. Children who cannot deal with their environments or demonstrate abilities that are reasonably expected of their age group are not ready for school. Discussions of competence and readiness in fact rely on the same themes. For example, the Head Start Bureau lists healthy growth and development, preschool education, and families who nurture their children's learning among objectives that support social competence. The three objectives for National Education Goal One (school readiness) are exactly the same. I shared these thoughts with Congress, which officially changed Head Start's goal to school readiness in the Coats Human Services Reauthorization Act of 1998. (Lest I be accused of having a change of heart, I will repeat that the planning committee was charged with designing an intervention to give poor children a "head start" so they would arrive at school with skills comparable to those of classmates from wealthier homes.)

Program Quality. My second line of defense for saving Head Start was to clean up program quality, which had been uneven since the program's hasty beginnings. I believed that because we were serving children from disadvantaged backgrounds, the program had to be of the highest quality to help them become ready for school. If we couldn't demonstrate such efficacy, I knew it would only be a matter of time before Head Start was deemed useless and closed down. Our only hope, then, was to improve quality and clearly demonstrate by school readiness outcomes that Head Start was an effective intervention.

I immediately initiated efforts to develop standards to govern the content and quality of Head Start services. It was then that I discovered how difficult it is to bring together politics, knowledge about child development, and common sense. There were committees, meetings, debates, and more meetings. The resulting Program Performance Standards were finally implemented in 1975, a full ten years after the project started. Model Head Start programs that served children younger than preschool age had no quality dictates at all until 1998, when the Performance Standards were revised and extended to cover the new Early Head Start.

The result of the lack of attention to quality over the years, including the lack of effective monitoring, was an erosion in the strength of the services delivered in some Head Start centers. Problems with quality were enunciated in the report of the Fifteenth Anniversary Committee, which I chaired, and repeated by its successor on the twenty-fifth anniversary, the Blue Ribbon Panel. Both reports were ignored, and the problems became worse in the early 1990s when a period of rapid expansion began. The growth caused many fragile programs to suffer further deterioration. I knew that many Head Start participants were not being served well, and I did not want to see more of them served that way. I therefore made a move I do not regret. On the basis of reports to me by old friends in the Head Start community, as well as some data at my disposal, I told the press that one-third of Head Start centers were of such poor quality they should be closed.

My friends in Head Start were shocked and hurt, and they were not shy about telling me so. But soon after the disgruntled phone calls and hate mail subsided, what I had tried to do so long ago in Washington actually came to be. The secretary of Health and Human Services, Donna Shalala, convened the Advisory Committee on Head Start Quality and Expansion. Congress passed legislation setting aside funds for quality improvements. Research Centers on Head Start Quality were formed, and measures to assess both the program's and participants' progress are being developed. Under the guidance of Olivia Golden and Helen Taylor, the performance standards were updated and strengthened. With Shalala's moral support, the Head Start Bureau has been able to defund over ninety grantees with incorrigible records of poor performance. When I was in Washington, I couldn't even close one. All these efforts have produced results, and Head Start is now better than it has ever been. It still has some way to go, but the strong commitment to improvement by the current leaders of Head Start and by Congress bodes well for further progress.

Experimentation. Another way I tried to strengthen Head Start when I was at OCD was to experiment with new methods of intervention. I envisioned Head Start as a national laboratory where we could develop new approaches to meeting the needs of children and families and refine existing ones. This way we could meet the critics by demonstrating that we were working hard to do better. I mounted such programs as Home Start, a home-based version of Head Start that still operates; the Child and Family Resource Program; and a national parent education program for adolescents. Some efforts worked, others failed, but at least we kept the experiment alive. Over the years, the many Head Start models and demonstrations developed in the national lab have added a great deal to the knowledge base in the fields of early childhood development, intervention, and family support.

Hopes and Disappointments

The sense of accomplishment derived from all these efforts has been more than offset by my failure to do more to assure good quality child care for every child who needs it. The national child care initiatives I envisioned for OCD never materialized. The planned child care system evaporated when Nixon vetoed the Comprehensive Child Development Act of 1971. That event remains the biggest disappointment of my career, and a monumental loss for generations of children. With the veto, we lost the opportunity to establish a government-subsidized national network of high-quality child care centers that would be accessible to all families because fees would be calibrated to income.

I began to search for other ways to meet our nation's need for high-quality child care that families can afford. In Washington, I developed the Child Development Associate program to train providers to give good care. The CDA has now been awarded to some 100,000 caregivers. The credential is nationally recognized and even required in some early childhood programs.

My first highly visible public effort after arriving in Washington was to implement the Arlie House Conference on Child Care. This conference assembled our nation's finest workers in child care. They presented a clear outline of what features must be present to assure good care for children of all ages. This effort was quickly followed by our construction of the first set of truly enforceable child care standards, called the Federal Interagency Day Care Requirements (FIDCR). My hope was that programs would have to meet these standards before they could receive federal funds. The National Association for the Education of Young Children currently promotes a revised version of the FIDCR. However, my twenty-five-year effort to make federal funding of child care contingent on meeting these standards has resulted in little more than a quarter of a century of failure. For nearly three decades, I've sadly witnessed our decision makers' willingness to watch tens of thousands of our children go to child care settings every day that are so poor in quality as to compromise healthy growth and development.

Although I haven't given up on standards, I tried another approach to our nation's child care dilemma. This was to tie child care closely to education and place it in the schools. Since I believe poor quality child care results in attenuated school performance, my own recommendations for school reform were presented more than ten years ago in my model of the school of the twenty-first century—the 21C schools—called family resource centers in Connecticut and Kentucky. There are now over six hundred of these schools in seventeen states. Kentucky has implemented this model statewide. A unit in Yale's Bush Center does nothing but help states and localities implement 21C schools. It took awhile, but the education establishment is now realizing that children's readiness and achievement are linked to the experiences they have before school age and after the school day.

Another way I approached the child care crisis was to attempt to reduce the need for supplementary care for those who have the hardest time finding it—parents of infants. Not only are an increasing number of mothers returning to the workforce during their infant's first year of life, the care they are finding is very expensive and often not very good. I worried so much about the developmental consequences to babies who are placed in child care soon after birth that I convened a panel of experts to study their caregiving needs and to design social policy to address them. The resulting proposal wandered through Congress for *eight years* before emerging as the Family and Medical Leave Act of 1993, which provides twelve weeks of unpaid leave. It was a far cry from the original, which had a longer paid leave, but it was a base on which to build. I have continued to push for a more generous leave, so recent proposals in this direction by policymakers and candidates are music to my ears. My colleagues and I have another ongoing project to analyze state child care regulations, especially those affecting infant care, and to make a case for their improvement.

As for the Children's Bureau, it was descending into what my late colleague Bill Kessen called "Federal invisibility" when I arrived in Washington, and there was nothing I could do to stop it. As the bureau's chief, I had two time-honored plans. One was to scan the social scene to discover the problems facing children and families. Solutions could be developed and tested in the national laboratory I was incorporating into Head Start. Second, I thought I was supposed to carry on the tradition of chiefs before me by highlighting the plight of children. The purpose was to make the public and the policymakers aware so they would rally around the cause. In appearances around the nation, I talked about children's suffering at the hands of poverty, abuse and neglect, and inadequate education, health, and child care services. My pulpit was quickly removed by a counselor to President Nixon. He told me that children could not possibly be faring so poorly while Richard Nixon was president. I am not quite sure if he truly believed that children were doing just fine under his president's care, but I am sure that I was supposed to stop talking about it. The bureau's mission of investigating and advocating for children's needs has now unofficially gone to the private sector.

On-the-Job Preparation

To achieve any of what I did accomplish in government, I had to fill in a huge void in my professional training. When I first came to Washington, I immediately realized how little I knew about the policymaking process. I also saw how little policymakers knew of the child development literature. They were very eager to learn. I came to be called "Professor Zigler," and my staff meetings were known as "lectures." This of course was not academia and the people who attended were not my staff or my students but very high-ranking officials. They were dedicated public workers who really wanted to hear what research could tell them, and to use this information to form better policies.

I never envisioned myself in this position, or in the federal administration at all. Recall that I began by describing myself as a basic scientist. My business was to produce and refine knowledge and theory. My work was shared with colleagues in professional journals and academic lecture halls. When I agreed to join Cooke's planning committee, my mentors were not particularly happy with me. At that time, basic and applied science were viewed as opposite ends of scholarship, and basic researchers as superior to those who worked in application. In government, I realized how artificial this split was. I had access to a wealth of knowledge derived from decades of work by basic scientists. Social policies were being formed without the benefit of this literature. Policymakers welcomed me as an interpreter and tried to act on what I taught them.

My colleagues in basic research were not quite so welcoming of the notion of attaching policy implications to their science. As an illustration of their stance, I vividly recall a meeting of the governing council of the Society for Research in Child Development (SRCD), a professional association dedicated to methodologically rigorous, theory-based studies of developmental phenomena. The time was 1975, when the United States withdrew from the conflict in Vietnam. Fearful of the future, Vietnamese parents were hoping to save their children by putting them on planes bound for the United States—a frantic effort dubbed "Operation Babylift." At the council meeting, I pointed out that our membership had a wealth of expertise that could aid in the resettlement of these children. We knew about the effects of social deprivation, about the potential developmental harm after separation from an attachment figure, about difficulties in adapting to a new environment. I suggested that SRCD write to President Ford and offer our services. SRCD's president said that he would send a letter offering his personal help, but he argued that the society should not be committed to such tasks. The council did see the babylift as a wonderful opportunity to study disrupted children. Instead of assisting the government, they decided to request government funding for this research.

Eventually the Ford administration asked me to convene a panel of experts to advise the administration on how to meet the needs of these children. The panel contained good representation by SRCD members, all of course acting on their own. Actually, by this time a growing number of us did believe that policy construction could be enriched by developmental science. Seeing how this could happen firsthand motivated me to join three leading developmentalists—Urie Bronfenbrenner, Julius Richmond, and Sheldon White—in approaching the Bush Foundation with a new concept. We proposed establishing centers for training child development scholars who wanted to work at the intersect of research and social policy. The Yale Bush Center in Child Development and Social Policy, which I still direct, has trained many fellows over the years to use their knowledge to solve social problems. Although mine is the last remaining center under the Bush name, there are many similar institutions today, including the one here at Fordham.

The mission of such centers has now become more acceptable in academic circles. Government and many private funding sources are seeking projects that have visible results. The spread of interdisciplinary work has brought psychologists closer to those in fields devoted to practice. Even SRCD has begun to welcome applied research into the pages of its flagship journal, *Child Development*. And today's young students are an active-minded generation who want to contribute both to the knowledge base and to humanity.

Conclusion

Thinking back, I attribute my success in having some influence over national social policies to my posture as an educator, not an advocate. As I learned over my time from planning Head Start to advising its administrators, developmentalists can accomplish much more by telling policymakers what we know and admitting what is not yet known, separating our facts from our opinions. I believe officials have listened to me over the years because they see me as a scholar, not as an advocate who has some other agenda. My only agenda, the one that has guided me throughout my career, is to serve the best interests of children and families.

EDWARD ZIGLER *is Sterling Professor of Psychology at Yale University and director, Bush Center in Child Development and Social Policy, Yale University, New Haven, Connecticut.*

SALLY J. STYFCO *is associate director, Head Start Section, Bush Center in Child Development and Social Policy, and research associate, Department of Psychology, Yale University, New Haven, Connecticut.*

2

A discussion of marginality and the variety of experiences that influenced a prominent developmental psychologist to forge a career in child advocacy.

Where Are You From? Child Advocacy and the Benefits of Marginality

Ruby Takanishi

As an applied developmental psychologist from long before the term became widely accepted, I have always been interested in life histories, which are the essence of the study of human development. Nevitt Sanford, one of my memorable undergraduate professors, taught me the value of the study of individual lives. What follows is a three-and-a-half-decade account of my professional journey to connect research about child and adolescent development to the formation of sound policies and programs that might contribute to increasing children's prospects for a good life. The fortuitous combination of historical events, mentors, and willingness to take risks to achieve clear goals, has resulted in my professional life history to this point.

Where Are You From?

I continue to be in places where it is unusual to have an Asian American woman of my generation or a person who looks like me sitting at the table or in the taxi. So even before the typical question, "What do you do?" I am often asked, "Where are you from?" or occasionally, "What are you?" Now the first challenge is trying to respond to what the questioner is really asking. I name the current city in which I am living, which is never satisfactory; but I have learned that the question is really, What country or ethnic group do you come from, and why are you here? I might feel cooperative and say: "Fourth-generation American of Japanese descent, born and raised in Hawaii." That usually satisfies the questioner, but it rarely deals with the second set of questions: Why are you here? And how did you get here?

Since these questions show no signs of ceasing, no matter where I am and how old I get, I am forced to reflect on what people are really asking. I must conclude that these are important questions. It means that I am noticed in certain ways, that I do not blend in easily wherever I am. And that is how I started to think about what it means to be a marginal person in a positive sense, and what the benefits of marginality are, especially in my line of work.

Marginality

I focus on the idea of marginality, because in my lifetime, those who work at the intersection of child development research and policy are still, in many ways, marginal men and women. There are exceptions, like my mentor Edward F. Zigler, who have attained professional and public recognition. Most academics who engage in applied and policy work are still considered strange fish. Those who are not in academe find themselves in a variety of settings—government, nonprofit organizations, and philanthropy. There, as in academe, we are different from those educated or prepared to work specifically in those sectors.

Many people tend to think of marginality as a negative quality. I am trying to put a positive spin on this asset of mine and how it plays into advocacy for children. The very task of connecting imperfect research knowledge with policy formation and program development requires a nonstandard set of skills and interests. It requires a person, ideally, to have mastered research knowledge and methods as well as the practical art of policies and programs. These are, as the philosopher scientist C. P. Snow observed, "two cultures."

Attempting such an integration quickly reveals the limits of our knowledge. As a researcher who adheres to certain standards and ethics, it will be difficult to make the concrete recommendations that are required by the policy and political process. As a researcher, you have to be smarter and more broadly trained than most others who remain within academic and research institutions.

Connecting, translating, and drawing concrete recommendations for policy and practice that are deeply respectful and understanding of the limits of research requires the best minds our field has to offer. My fundamental, most critical requirement for connecting research with policy is to obtain the best, most rigorous training in the methods and approaches of the disciplines that you can.

Being an advocate for something means that you have to believe strongly in what you are supporting. But for those who are scientifically trained, this stance goes against a core value—being constantly skeptical. Marginality in connecting science and policy is, interestingly, a fundamental criterion for effectiveness in this line of work. You can never really be a true believer, and thus you will remain marginal—but it is critical to develop a sense of comfort with your marginality, because this work is a long-term commitment that requires persistence and focus in the context of setbacks and disappointments. Good outcomes are often opposed for no good reason. No dilettantes or control freaks need apply.

I am the only person among the distinguished psychologists in this volume who is not based in academe. I point to this fact because being a professor has a recognized place and reality for most people. But increasingly, individuals who connect research with policy will find themselves elsewhere, in research organizations in the nonprofit and for-profit sectors, in media, child advocacy organizations, philanthropic organizations, government at different levels, and service organizations both private and public.

My own choices reflect a brief period in the academy from 1973 to 1981, but since I was a Congressional Science Fellow in 1980–81, I have had the good fortune to spend time in various settings that have provided opportunities to connect my research interests with policy and program development. But why and how did I get there?

Crucial Beginnings

Parents matter, and in my case, they still matter a lot. My beginnings, especially influenced by my mother, reflect a high value on science and an equally high value on service or "giving back." I was raised with the idea that those who are fortunate are obligated (in the Japanese tradition of *giri*) to reciprocate, not only to their elders but also to future generations. My grandparents and preceding generations believed that they were working toward a better future for their children, and they were willing to make great sacrifices to assure that future. The Japanese Americans have a saying: *kodomo no tame ni*—for the sake of the children. This was the guiding principle in my formative years and continues to guide me as an adult.

Among the immigrant workers in the small sugar plantation town in which I grew up, knowledge and skills had to be useful; they had to improve conditions and the lives of individuals. I attribute my lifelong interest in how inquiry, research, and knowledge can inform policy and ultimately life prospects to these core values.

As Glen Elder (1999) documented so well in his classic study *Children of the Great Depression*, the specific historical period for development during young adulthood is strongly formative. To have come of age in the mid-sixties, and to have selected developmental psychology as a field of study exactly as the first summer Head Start programs in 1965 were initiated, was to participate in the excitement that the very field I chose could contribute to reducing poverty and improving the life prospects of children in America.

I remember visiting the Perry Preschool Project in Ypsilanti, Michigan, in the summer of 1970. (This project, also known as High/Scope Educational Research Foundation, followed 123 low-income children from the time of their participation through the age of twenty-seven.) Little did I realize then that the longitudinal research findings of the Perry Preschool would be the major research rationale for early childhood programs in the next three decades. I remember the agonies of the Westinghouse evaluation of Head Start (Cicirelli, 1969), and the struggles to keep that program alive as Southern congressmen railed against local control. I remember how profoundly discouraged I was when I learned of President Nixon's veto in 1971 of the proposed Comprehensive Child Development Act, which would have created a universal system of early childhood education and care for all children. And I remember the growing disillusionment during the seventies regarding our capacity to contribute to the solution of social problems. I came of age during an era of great hopes, but have lived my adult years with the constant erosion of those hopes. To have started out as a young person during Great Society days and to have lived my adulthood years with small victories has constantly tested my endurance.

I completed graduate school in June 1973, when academic opportunities were not ample, and I was neither a real developmental psychologist nor a real early educator. I had experience in both fields, saw myself as belonging to both, but pretty much fell through the cracks.

Fate was kind, and I managed to land in a tenure-track position at UCLA's Department of Education, where Norma Feshbach had put together a National Institute of Mental Health (NIMH) training grant to prepare young people to combine their interests in child development with the education of children. She was an early practitioner of what is now called applied developmental psychology. It was there that I found my academic home, creating and teaching courses in research and evaluation of early childhood programs, and in child development and social policy.

Once again, historical events intervened when foundations like the William T. Grant Foundation and the Foundation for Child Development initiated the Congressional Science Fellowship Program in 1978, and the Bush Foundation (Minneapolis) launched the Bush Centers in Child Development and Social Policy. I was able to contribute to the shaping of the Bush Center at UCLA the summer before my postdoctoral year at Harvard University (1978–79).

Auditing science policy seminars at the Harvard John F. Kennedy School of Public Policy introduced me to the politics of how a society allocates public support for research and development. I also learned that the distinguished professors at the Kennedy School had no idea that developmental psychology existed, and that support of the behavioral and social sciences was a matter of serious debate. Since I was at that time an adult in my thirties, my experiences with those nationally recognized professors reinforced my experience that the nexus of being Asian American, a woman, and an applied developmental psychologist interested in policies affecting children and families is one of extreme marginality, and in some cases of outright ridicule. These experiences constituted a reality check on the status of an enterprise that we insiders often take for granted and often feel quite satisfied about.

Taking Risks

Tenure at UCLA coincided with an award of a Society for Research in Child Development Congressional Science Fellowship to work in the U.S. Congress in the fall of 1980. It was my first or second day on the job on Capitol Hill working for the senator from Hawaii when the dean of my school gave me a career-changing choice—to come back to UCLA after the two-year leave that had already occurred, or to sever my ties to UCLA and likely to a university career. I said that I was beginning to do just what I wanted to do, and I was not coming back.

With President Jimmy Carter's early acceptance of defeat by Ronald Reagan, the leadership of the U.S. Senate shifted overnight from Democratic to Republican and I became a minority staff member. When Ronald Reagan became president in 1981, the historic beginning of what we now call devolution began with the block granting or consolidation of federal programs focused in special areas, the conservative attacks on the value of "social research," and efforts to dismantle the statistical systems of the United States. The Conservative Revolution had begun.

I know now what people mean when they say they witnessed the making of history. Being a legislative assistant at the beginning of the Reagan administration—particularly for social and children's programs and the funding of behavioral and social science research—was to see the inception of historical changes that persisted through the next two decades. The Personal Responsibility and Work Opportunity Reconciliation Act of 1996, referred to as "welfare reform," was its intended outcome, and took fifteen years to become law.

I have few regrets about my decision in September 1980. After my brief time in the academy, I was ready to move on and take a more active, engaged role in the issues that really mattered to me.

I learned during my Congressional Fellowship what it is like to be on the receiving side of advocacy. As children's advocates came to make their case for funding of their programs, such as Head Start and foster care, I quickly learned that moral arguments have their limits. What an elected official wants to know is what specifically should be done within existing or proposed legislation and what the political support for the proposals are. What is good for children has to compete with many other interests, some good and some wasteful, especially in the appropriations process.

I learned the art of negotiation and compromise, of incremental change, of losing gracefully and not making enemies, and of being tenacious over the years. Most important, I learned that relationships with key staff and building trust and credibility over time matter a great deal. Now, even more than twenty years later, there are so many competing interests, so many competing sources of information, and so little time to make decisions that may affect the lives of many. To have any hope of being effective, a person must be available, ready to respond quickly and briefly, and value the dance of legislation.

Research is part of this process, but it is only a very small part. Sometimes research seems to make a big difference, but these "big bang" examples are rare. When you examine the situation, being in the right place at the right time and having the necessary relationships can make all the difference. It is not only knowing relevant research findings, but knowing when and how to use them to inform policy, and typically as quickly as possible. Good, clear communications skills and persuasiveness are essential.

Taking the risk in giving up tenured lifetime employment at a good university in 1980 may have been foolhardy. People still gasp at the audacity or the stupidity of what I did. I do believe that it is important to know when to take such risks when you have a goal in mind. This is not easy to do when one has to feed oneself and have a roof over one's head, and a supportive spouse and family should not be taken for granted.

But what taking risks does mean is that many times you have to be willing to go places and take positions that are not going to result in lifetime employment, and may have uncertain outcomes from the beginning. I am glad I read Seymour Sarason's books (1972, 1977) in graduate school, in which he wrote about adults' having several jobs over their lifetimes. I accepted his prediction wholeheartedly; however, in my later years, I have come to appreciate the value of stability in work and place.

Life Outside Academe

After my Congressional Fellowship in 1981, I returned to New York City where I took temporary teaching positions at Teachers College, Columbia University, and at the Bank Street College of Education. Edward Zigler honored me by asking me to teach his graduate seminar in child development and social policy at Yale while he was on sabbatical.

In 1982, I returned to Washington to take two positions that drew on my postdoctoral experience, where I had focused on science policy and children's issues, and on my Congressional Fellowship experience. I became the founding executive director of the Federation of Behavioral, Psychological, and Cognitive Sciences, a new cluster of small scientific societies interested in banding together to defend their funding and science policy. I was involved in organizing the first Congressional Science Seminars, which continue today, for Congressional staff to hear from leading behavioral and social scientists about research enterprises and findings relevant to salient policy issues. I also organized the Science Policy Seminars, which also continue, in which government science administrators and bench scientists meet to discuss policies that affect the conduct of research, particularly as funded by the public sector.

During the same time, I became a registered lobbyist for the American Psychological Association Office of Public Policy, working for children and family issues. In 1984 I became head of the APA Office of Scientific Affairs, where my responsibilities included revision of standards for psychological and educational testing, ethical guidelines for research with animals and with human participants, scientific awards, and science policy, specifically funding of the behavioral and social science enterprise. I took this administrative position because I wanted some significant managerial experience.

At APA I learned that the research enterprise is not a clean and linear process. It is fashioned from competing interest groups and egos, just like the U.S. Senate, which I had just left! But as history would have it, my two years as head of the APA Office of Scientific Affairs coincided with its splintering into APA and the American Psychological Society (APS). I am proud of initiating and implementing the chief scientific adviser position of the APA, and as fate would again have it, to leave before its formal dissolution.

It was from that position that I was unexpectedly asked to head the Carnegie Council on Adolescent Development in the summer of 1986. It was not a position to which I applied, nor was adolescent development an area in which I had any expertise. But the combination of research training and science policy, a steady commitment to children, and academic, legislative, and nonprofit organization experience, as well as administrative and management skills, was probably appealing to those who were starting this brand new enterprise.

A Decade with Carnegie Corporation of New York

When I started my work as the founding executive director of the Carnegie Council on Adolescent Development, an operating program within a national foundation, I had no experience in philanthropy. I was asked to start from scratch, a new program to raise adolescents higher on the national agenda funded as an internal program of a distinguished national foundation. From the first day, I felt the heavy responsibility to justify the value of an internal foundation program in an organization that typically provides funds to outside groups whose existence often depends on foundation grants.

David A. Hamburg, president emeritus of Carnegie Corporation of New York and my boss for ten years, shared my professional passion for connecting research with public policy. This was a productive convergence of interests. Every initiative of the Carnegie Council, from its landmark report on the reform of middle grades education, *Turning Points* (Task Force on Education of Young Adolescents, 1989), to its collaboration with the former U.S. Congress Office of Technology Assessment (OTA) on the health status of American adolescents and its timely analysis of the need for after-school activities represented by *A Matter of Time* (Task Force on Youth Development and Community Programs, 1992), was based firmly in the current research.

As part of these activities, research was commissioned and stimulated by a series of science policy conferences attended by both public and private funders. These activities contributed to the creation of a U.S. Office of Adolescent Health within the Maternal and Child Health Division of the U.S. Department of Health and Human Services, and to new funding for adolescent health research in the U.S. National Institutes of Health. *A Matter of Time* contributed to crime prevention programs and to federal funding of the 21C community learning schools, and rode the crest of interest in how after-school programs kept young children and adolescents safe during the after-school hours. In retrospect, it is remarkable that since the publication of that report in December 1992, after-school programs have gained wide acceptance and sizable funding increases, whereas before they had not been on the public agenda.

At the end of ten years, the council was deemed a successful initiative that contributed to elevating the second decade of life on the national agenda. To be part of building that initiative, especially its strategic aspects that rested on an institutional approach—schools, health organizations, community organizations, media, and families, all with a science base—was a satisfying experience. The growing interest in the public and private sector on youth development is partially due to the work of the council. The Carnegie Council effort connected research to policy and program development in every aspect of its work, and that remains its distinguishing legacy.

Public Service: A Brief Stint in the President's Science Office

When David Hamburg decided to retire from the presidency of Carnegie Corporation of New York, it was time for me to seek new venues. At that same time, I was asked to join the staff of the President's Office of Science and Technology Policy (OSTP) as the assistant director for behavioral and social sciences and education. What attracted me to brief public service was an idea of the OSTP associate director for science, physicist Ernest Moniz from the Massachusetts Institute for Technology, that it would be useful to estimate the level of federal expenditures for research and development on children and adolescents, and to convene all of the federal agencies supporting research in this area to identify research initiatives that could be linked to the Domestic Policy Council in the White House. Moniz had the right idea. Policymakers had no sense of what the federal agencies were investing in the first two decades of life. By contrast, estimates for space exploration, cancer, education, and technology were widely available. What percentage did this investment represent in terms of the federal investment in children and families? But the bottom line was that Moniz had the vision that this kind of information and its yield could be more closely connected with how domestic policy was shaped at the national level. After all, developments in physics were systematically tied to their applications in various fields of engineering and technology.

A person was needed who knew federal agencies, science support policy, and the politics of working in the agency environment, and who had the proper political credentials. I fit the bill on most accounts. In the tradition of testing the new person, my first assignment was to take a 740-page report on regulations regarding advertising cigarettes to children and adolescents submitted by the Food and Drug Administration to the U.S. Office of Management and Budget, and comment on it for OSTP Director John Gibbons—in one hour! I told them that I would be happy to provide it the next morning, and I did, bringing to bear research on the factors that contribute to smoking among youth, and the limits of the assembled knowledge to inform policy.

My OSTP experience informed and reinforced some conclusions I had come to over the years.

The first is that the scientific disciplines that are well organized and familiar with making their case to the federal government on a regular basis were the ones who were most likely to get funding. I was shocked by the lack of initiatives by the social and behavioral sciences associations.

The second is that there is little incentive for federal agencies to work together, especially when it means that they will lose control over their programs and face the possibility of future lower funding levels even though the benefits may be enhanced visibility and prestige. However, higher-level entities such as the President's Council of Scientific Advisors and other leaders can play an important role in directing federal agencies to work together on areas of mutual interest, such as how children learn and the role of current and emerging technologies. Working together creates an innovative push against natural agency inertia. Such a council, by advising the president, could provide stimulus for the Domestic Policy Council to pay attention to certain neglected issues.

I obtained an excellent overview of how science policy is shaped at the federal and agency levels. The analogy to sausage making is not far from the truth! But individuals with vision and the willingness to stake their careers on certain issues can make a big difference. And I learned that, contrary to prevailing views about government bureaucracies, the Science Division of OSTP during the time I was there had both civil servants and temporaries like me who really cared, worked really hard, tried to make a difference, and were as outstanding a group of committed colleagues as I have had in any position.

My stint of six months coincided with the presidential election of 1996. Election years pose difficult assignments, and I had to defend national education standards about which I had doubts; I lost battles, such as pushing for universal preschool, because they were considered too expensive before the time of budget "surpluses." Four years later, in the presidential campaign of 2000, universal preschool and educational Head Start programs were prominent on the education agendas of both candidates, given the apparent surplus. This taught me the importance of taking a long-range view, living with political considerations or imperatives, and not giving up on ideas that are important. Endurance, long-term perspective, and building constituencies continue to be necessary to achieve goals for policy change and initiatives.

The Latest Chapter

In December 1996, I became the fifth president of Foundation for Child Development (FCD), a national privately endowed philanthropy with a century of commitment to children and their families. Once again I was living in interesting times; the period of the nineties was one of unprecedented prosperity for the United States and certainly in the growth of philanthropic assets. This very fact put the spotlight on this sector and created opportunities for the diversification of philanthropy, particularly new forms of venture philanthropy and ideas of social entrepreneurs. In brief, these forms of philanthropy practiced by young, newly wealthy individuals draw from corporate practices of business plans, have close involvement of the benefactors in implementation of these plans, and reflect their insistence on concrete, measurable outcomes. These developments have affected more traditional foundations like my own, and specifically for FCD, raised questions about how relatively small amounts of money can be granted strategically to affect broader public policies.

As a small foundation, FCD supports research, policy analysis and development, and advocacy on behalf of low-income working families. At the root of all our efforts is our research base of knowledge, which comes from FCD's origins in the Progressive Era. In brief, the Progressive Era pioneered the use of social surveys and other forms of data collection as well as demonstration projects to improve conditions for children and families. This is one of the roots of the current interest in linking research knowledge about children with developing social policies related to them.

It is premature to assess what this phase of my professional life will yield, but once again, I find myself in a place where the opportunities to connect and fund research to policymaking are ample. The most challenging task of my presidency thus far is how to sustain a pipeline of individuals who are interested in connecting research with social policies concerning children as the previous generation of scholars and practitioners funded by FCD, the William T. Grant Foundation, and others, begin their retirement years. Some aspects of this pipeline, fortunately, have been institutionalized. The Society for Research in Child Development (SRCD) Congressional Fellowship Program has been revived. SRCD's executive fellowship program continues. A number of graduate schools of public policy are producing individuals who have earned a public policy degree combined with a Ph.D. in a behavioral or social science. With the massive retirements of the baby boomers, key agencies like the Congressional Research Service will require significant numbers of new recruits. As a foundation, FCD continues to seek opportunities to seed and to support efforts that increase the numbers of individuals prepared for these roles.

Where Are You From? Child Advocacy and the Benefits of Marginality

I return to my starting point. My professional history reflects the positive value of marginality. In almost every organizational or institutional setting in which I have worked, I was not part of that world before, nor was I specifically prepared for that work. I was an outsider, and even after years in that setting, remained a skeptical outsider. The scientific values of skepticism and incomplete knowledge that I accepted at an early age continue to be guiding principles.

Marginality is an asset because it puts you in a position to question and not to accept the prevailing ideas or wisdom. You can think about not what is but what could be. As a perennial skeptic, you are always questioning and looking for other ways, especially valuable when prospecting for new ideas and good people.

I have been very fortunate. I have had wonderful mentors and opportunities. It's not that I haven't worked hard. But the fact of the matter is that there are a lot of people with similar or better skills who never had the opportunities I had to do what I set out to do. So what I have tried to do is to give back—to increase the opportunities for other individuals to do the kind of work that I do. We want all young people who choose to have influential lives to succeed, because many more are needed to assure that all our children have better futures than they now do.

Four ingredients have contributed to my life history thus far: a high level of research training and thorough socialization about the importance of scientific inquiry; a focused search for knowledge and skills outside traditional graduate education, including internships and coursework in public policy; a group of outstanding mentors who have continued to stand by me throughout my professional life; and a willingness and need to take reasonable risks in moving from one organizational setting to another. I share these ingredients because they are essential to individuals who are seriously considering a professional life that aims to connect research about human development with policy.

Most important, all of us, including those of us who have "been there," must take a long-term perspective and accept that we may not get to see what we really want to see happen. I was reminded of this recently with Ken Burns's documentary (2000) on the life of Elizabeth Cady Stanton and the struggle for women's suffrage. I was reminded of this when Brian O'Connell

(1999) described the fifty-year struggle for acceptance of universal kindergarten.

I like to think that some of us have made a clearing for those who will follow. While it may be fun to have been a pioneer, it is much more important to increase numbers who will settle in and cultivate this territory. Bon courage. Bon voyage!

Acknowledgments

I thank Joanne Mackie and Sara Vecchiotti for their assistance in finalizing this chapter.

References

Burns, K. (director-producer), and Barnes, P. (producer). *Not for Ourselves Alone: The Story of Elizabeth Cady Stanton and Susan B. Anthony*. 2000. Film. (Available from Public Broadcasting System, http://www.pbs.org/).

Cicirelli, V. G. *The Impact of Head Start: An Evaluation of the Effects of Head Start on Children's Cognitive and Affective Development*. Washington, D.C.: Westinghouse Learning Corporation, 1969.

Comprehensive Child Development Act of 1971, H.R. 6748, 92d Cong., 1st session, 1971.

Elder, G. *Children of the Great Depression: Social Change in Life Experience*. (25th anniversary ed.) Boulder, Colo.: Westview Press, 1999.

O'Connell, B. *Civil Society: The Underpinnings of American Democracy*. Hanover, N.H.: University Press of New England, 1999.

Sarason, S. B. *The Creation of Settings and the Future Societies*. San Francisco: Jossey-Bass, 1972.

Sarason, S. B. *Human Services and Resource Networks*. San Francisco: Jossey-Bass, 1972.

Task Force on the Education of Young Adolescents, Carnegie Council on Adolescent Development. *Turning Points: Preparing American Youth for the Twenty-First Century*. New York: Carnegie Corporation of New York, 1989.

Task Force on Youth Development and Community Programs, Carnegie Council on Adolescent Development. *A Matter of Time: Risk and Opportunity in the Nonschool Hours*. New York: Carnegie Council on Adolescent Development, 1992.

RUBY TAKANISHI is president of the Foundation for Child Development.

3

Research can contribute to policy by asking good questions, using the best methods we have, and being ready when the policy issues arise.

From Research to Policy: Choosing Questions and Interpreting the Answers

Aletha C. Huston

Academics are not accustomed to putting their personal lives in print, so this assignment to tell my life story is a difficult one. I will outline some themes and principles that have guided my efforts, and describe some of the conclusions I have drawn about how to increase the likelihood that one's research will have an impact on policy. Certainly, this after-the-fact summary sounds more coherent and planful than it should. Lives and careers are strongly affected by unpredictable events and serendipity. Part of the story is balancing your chosen directions with changing opportunities and social issues, having some goals but being alert to opportunities that arise.

Because this is a life story and because we developmental psychologists believe that childhood experiences are formative, I'll start with some themes from my early years that have guided me throughout my professional life. My parents were strong New Deal Democrats, devoted followers of Franklin Roosevelt in the 1930s and 1940s. My father ran for Congress in 1940 and 1942 on the Democratic ticket in downstate Illinois, where Democrats were in the minority. To my mother and father, Roosevelt's New Deal represented the philosophy that government can improve the lives of the people who are its constituents, that government should act to assure fairness, equity, and survival for everyone, and that government service, especially in an elected office, is a high calling. In 1944, my father was killed in the Second World War, and my mother forged a new life for herself. She became a school psychologist and, later, a professor of special education.

I grew up with the strong expectation that I would have a profession, a somewhat unusual circumstance for girls in the 1950s, and that my work should contribute to human welfare. These basic principles, that one should be of use and that government is an important arena for contributions to human well-being, have been anchor points in many decisions and choices throughout my professional life.

My professional paths were also influenced by my training. As an undergraduate at Stanford, I was given a heavy dose of basic psychology and the opportunity to do Albert Bandura's first experimental study of imitation as my undergraduate honors thesis (Bandura and Huston, 1961). Despite this experience, I thought my interests and abilities lay in applied work, and I got my Ph.D. in clinical child psychology. By the time I finished clinical training, I was frustrated with the ineffectiveness of available psychotherapeutic methods and welcomed the opportunity to pursue an academic research career in which I hoped to produce knowledge that had some social application.

Dimensions of Research

People often pose basic versus applied research as a dichotomy, but I find it more useful to think about this distinction as a continuum. The likely influence of a particular program of research varies along a dimension from indirect (basic) to direct (applied) and often rests on how research questions are formulated. There is a famous, though perhaps apocryphal story that, as Gertrude Stein lay on her deathbed, her followers eagerly asked her "What is the answer?" Her reply was "What is the question?" The questions in research determine the answers they will produce. Are questions drawn from theory or are they crafted to answer an applied question? Are they formulated in a way that will illuminate a choice among policies or practices?

On the continuum in Figure 3.1, the questions at the indirect end are generated from basic theory or curiosity. For example, Bandura's early research on imitation was designed to test hypotheses drawn from social learning theory about whether observational learning could occur without reinforcement and, once that was demonstrated, what processes of attention, memory, and motivation accounted for observational learning and imitative behavior (Bandura, 1969). At the direct end of the continuum, questions concern policy and practice. For example, researchers have asked whether students learn from such particular educational television programs as "Channel One News" when they are shown in the classroom (Johnston, Brzezinski, and Anderman, 1994).

The second continuum shown in Figure 3.1 is the level of analysis, ranging from the individual at one end to society and large aggregate units, including government, at the other. As psychologists we study individuals. Our applied research tends to investigate interventions and processes at the individual level (such as psychotherapy or parent training). Public policy

Figure 3.1. Two Research Continua

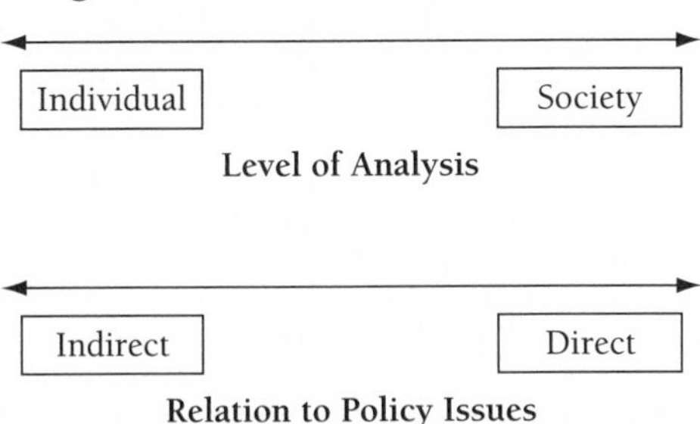

occurs at the aggregate level, and much of the influential research and scholarship comes from economics, sociology, and political science. My own work illustrates this continuum, having moved from the individual toward the aggregate.

Forming Questions

To inform policy or practice, issues of importance in those arenas ought to form our questions. At the indirect end of this continuum research can be designed to inform important social issues. For example, from 1965 until the mid-1980s I studied sex typing in children. During that time, the women's movement brought about a major shift in theory and assumptions about the value of conventionally masculine and feminine identities and behaviors. It seemed to me that studying sex differences simply fueled rather pointless debates about nature-nurture issues. Research to understand predictors of within-sex variations in cognitive or social characteristics was, in my opinion, more scientifically informative, and also more likely to lend itself to applications that might be helpful to children. That perspective led to research on the antecedents of achievement behavior in females (Stein and Bailey, 1973) and on the life-span development of women (Huston-Stein and Higgins-Trenk, 1978).

There is no formula for selecting questions, but here are a few guidelines. First, it is absolutely necessary that you care about the issue—that it resonates in your gut as well as your head. Research involves long, tedious hours, so I tell my students they should be passionate about their research topics.

Second, it matters that other people care about the issue, including policymakers and funders. Some of my long-standing research activities, such as my work on television and on child care influences, were initiated in response to National Institutes of Health funding programs. I do not recommend simply bending to the latest Request for Applications, but you can be primed to respond to an initiative that fits your interests and goals.

Third, does the question lend itself to solid empirical study? Is there a body of theory and data that can aid in conceptualizing it? Are there measures? Are strong methods possible? Can one design and carry out research that will do a good methodological job of answering the question—and will the answer have some generality and scientific value beyond the specific case addressed in the research? It is not very interesting to show, for example, that a particular program works unless you have some information that helps to understand what features of it were successful and may be generalized or adapted in other programs to increase their success. It is even less informative to show a program does not work without data to help understand why it failed.

Fourth, does the research generate information that could inform policy and practice if the opportunity arises? One classic concept in policy studies is the window of opportunity, a time when a topic reaches the top of the policy agenda and both policymakers and the public are receptive to information and potential solutions. Because good research takes time, policy researchers need to plan their research around issues that are likely to have enduring importance, and they should be ready to produce that research when the window opens. Most important issues do resurface periodically. For example, television violence has appeared repeatedly on the policy agenda since 1954 when Senator Estes Kefaufer, chair of the Senate Subcommittee on Juvenile Delinquency, conducted hearings questioning the need for violent content (Liebert, Sprafkin, and Davidson, 1982). With each new round of public and governmental concern, policymakers reexamine the accumulated research funded because agencies and researchers thought the topic was important.

Examples from My Research

In 1969, the National Institute of Mental Health (NIMH) initiated a program of research on television violence and aggression in conjunction with the formation of the Surgeon General's Scientific Advisory Committee on Television and Social Behavior. The major question framed by the funding agency was "Does viewing television violence affect real-world aggressive behavior?"

At the time, I was a young assistant professor collaborating with a colleague on soft money. We thought the topic was important, that our research was likely to be funded, and that findings would be highly visible to policymakers as part of the Surgeon General's report. However, we expanded the question to ask about the effects of prosocial television on prosocial behavior as well as the effects of violent television on aggression in order to test the general proposition that children learn and imitate a range of social behaviors portrayed on television. This gave the research more theoretical generality and widened the implications for media policy. If we found positive effects from prosocial programming (in this case, "Mister Rogers' Neighborhood"), it might suggest to television producers what to do as well as what *not* to do.

We based the study on Bandura's (1969) theory of observational learning. He conceptualized imitative behavior as the endpoint in a sequence that included attention, learning, and individual motivation. Hence we measured attention, recall, and children's initial behavioral dispositions to be aggressive and prosocial. Measuring the processes that were thought to mediate the influence of watching television on behavior provided some assurance that, if we did not see effects on behavior, we could understand why.

The study was a field experiment in which approximately a hundred children in four preschool classes were randomly assigned to see aggressive cartoons (violent TV), "Mister Rogers'" programs (prosocial TV), or neutral children's films with little social behavior (such as visits to the circus). The children were recruited in central Pennsylvania for a free summer preschool. They came from families across a wide range of socioeconomic status; they were almost all European American. Observers coded children's aggressive and prosocial behavior in the classroom. After a baseline period that allowed us to establish initial individual differences, children saw their assigned programs every school day over the next four weeks (Friedrich and Stein, 1973; Stein and Friedrich, 1975).

An experimental design was necessary to establish a causal relation between viewing and behavior. Conducting the experiment in preschool classrooms provided a good sample of natural real-world behavior. If there were effects, they should be persuasive, but this design also ran the risk that weak effects would not be evident in the complex setting of everyday play.

As it turned out, we did find some effects of exposure to both the aggressive and prosocial television programs. The study was widely cited, and it led us to a program of research pursuing the potential of prosocial programming for enhancing children's prosocial behavior. Some years later, my colleague John Wright and I conducted a series of studies showing the relations of viewing educational programs, particularly "Sesame Street," to children's cognitive development and school readiness (Huston and Wright, 1996, 1998; Wright and Huston, 1995; Wright and others, 2001).

I devoted my energies to studying the potential of television to benefit children rather than pursuing the effects of television violence because I thought there was more chance of influencing media policy to provide good television than there was of influencing it to decrease violent content.

This long program of research has had some impact on children's television policy and practice. In 1972, using the body of research generated by the NIMH program, the Surgeon General's Committee stated that television violence has a causal relation to aggressive behavior, at least for some people some of the time (Surgeon General's Scientific Advisory Committee on Television and Social Behavior, 1972).

Many professional and advocacy groups have since reached the same conclusion on the basis of available research. There have been periodic efforts by the U.S. Congress to generate legislation and by some presidents to persuade the television industry to reduce children's exposure to television violence. The most recent of these is the V-chip, which permits parents to block transmission of particular programs deemed inappropriate for children (Kunkel, 1998). In my opinion, however, we have made little headway dealing with the overall problem, in part because there is little that government can do to counteract the powerful economic and social forces supporting violence in the media.

Research on prosocial and educational television has made a more tangible contribution, but only when the political and social climate of the country was receptive. In the 1970s, both commercial and public television produced new positive programming for children. One major impetus for this trend was the spectacular success of "Sesame Street," and the documentation of its effects in a major evaluation of its first two seasons (Bogatz and Ball, 1972). Public funding for children's television increased, and commercial producers also responded. About 1974, CBS ran a large advertisement in the *New York Times* promoting its new "prosocial" programs, including such excellent shows as "Fat Albert and the Cosby Kids." After 1980, there was an era of deregulation, which led to a serious decline in the number of good children's programs and an increase in product-oriented programming for children—what some have called thirty-minute commercials. In 1990, the Children's Educational Television Act requiring broadcasters to serve the educational and informational needs of children became law. In 1996, the Federal Communications Commission (FCC) completed a rule-making process that made the requirement more specific at three hours per week (Kunkel, 1998).

What role did research play in these policy changes? And to what extent was my research planned to contribute to policy discussions? In my opinion, research by itself rarely brings about major changes in social policy or public opinion. Both government and industry policies are driven by strong political and economic forces. When research does not support people's ideological or economic goals, they typically ignore or discount it. When it supports their goals, however, it can provide some moderately powerful ammunition contributing to adoption of policies intended to achieve those goals. I testified in several congressional hearings during the late 1980s, always for a committee whose chair was promoting the policy supported by my research. In one hearing, the committee staff had assembled two large tables piled with scholarly volumes containing research on the effects of television violence on children and adolescents.

Although the Children's Television Act was passed in 1990, President George Bush's administration had little interest in enforcing it. President Clinton, by contrast, appointed an activist chair of the FCC who undertook a rule-making process to put some teeth into the vaguely worded act. Researchers, including John Wright and me, were invited by the White House and Vice President Gore to highly publicized conferences and media events in which we presented our research and its conclusions. The staff of the FCC did a thorough review of existing research, as they had done in the late 1970s. They interviewed us and many others about our research and opinions. Advocacy groups such as Action for Children's Television used our research. It was a bit disconcerting to hear on PBS pledge drives that viewers should support the station because "researchers at the University of Kansas" (where Wright and I conducted our research) had shown that "Sesame Street" increased school readiness. It is doubtful that the three-hour rule would have been put in place without supporting research, but also highly unlikely that research alone would have led to such a policy. In short, good research supporting a policy is a necessary but not sufficient condition.

This synopsis of thirty years of research on media represents one style of trying to ask questions and produce knowledge with social relevance. I chose questions that had social importance and tried to generate good research. I was not, however, an activist whose primary goal was to affect media policy. My colleagues and I published in conventional scholarly outlets, and we cooperated with others who translated our work into a form for a wider audience. When invited, we testified, summarized, talked to the media, and drew conclusions.

Children and Poverty

In the last fifteen years, I have undertaken another line of research dealing with children and poverty that falls closer to the direct end of the continuum in Figure 3.1. In the mid-1980s, the mass media began to publicize the fact that the percentage of children living in poverty had risen from about 14 percent in 1970 to over 20 percent in 1985. It was a discouraging time for many of us who had thought the War on Poverty and other social programs of the Johnson era would have lasting effects. As a citizen, I was troubled by the direction our society was taking, but I did not see a way to incorporate that concern into my research. Then I attended a lecture by Edward Zigler, and the proverbial light bulb went on. Public policy could be a legitimate subject for scholarly study, not just something that could be influenced indirectly by relevant data.

To pursue this new direction, I took advantage of a wonderful internal sabbatical program at the University of Kansas to study policy analysis in the Political Science Department. Like many ventures outside one's own culture, the most enlightening part of this experience was learning that policy analysts did not share many assumptions of psychologists. First, psychologists tend to assume that if we can show that something is beneficial for children, policymakers will surely want to adopt it. Policy analysis, however, is heavily influenced by economics, where potential benefits will always be weighed against costs. Second, psychologists conceptualize human behavior at the individual level, while political scientists, economists, and sociologists conceptualize it at the societal level. Finally, although children are central to developmentalists, they have relatively little importance in most other disciplines that influence public policy.

I concluded that to play in the policy game, research must be multidisciplinary, and that it is especially important to incorporate economic analyses. As a beginning, I organized a small working conference on children and poverty, with participants from sociology, economics, health, education, and psychology. The rich interchange led to an edited book, *Children in Poverty* (Huston, 1991).

Welfare and Antipoverty Policy. Welfare and antipoverty policies for adults are important potential influences on children's poverty and on the well-being of children living in poor families. In the late 1960s and early 1970s, the federal government had sponsored several experiments using random assignment methods to test the effects of a "negative income tax" or guaranteed minimum income (Salkind and Haskins, 1982). By the early 1980s, policymakers began experimenting with programs to help recipients of Aid to Families with Dependent Children (AFDC) become self-supporting. The Department of Health and Human Services, which oversaw AFDC, required high-quality evaluations of these trials, usually with random assignment experimental designs (Greenberg and Wiseman, 1992; Gueron and Pauly, 1991). These studies had a direct influence on federal legislation and government policies affecting poor families.

I was struck, however, with the fact that the indices of success were typically limited to economic indicators—changes in income and employment for parents and reduction in welfare costs. Even though mothers of children as young as one year of age could be required to participate in training and job searches, there was little or no information about the consequences for the children in affected families. Researchers at Child Trends in Washington had the same concerns and initiated some very important studies (for example, Zaslow, Moore, Morrison, and Coiro, 1995).

The New Hope Project. In 1994, the MacArthur Foundation formed the multidisciplinary Research Network on Successful Pathways Through Middle Childhood, which included Robert Granger from Manpower Demonstration Research Corporation (MDRC), the organization conducting many of the experiments on welfare reform. Granger invited the network to design "child and family" supplements for some of their ongoing studies. From that collaboration grew an investigation of the New Hope Project, an antipoverty initiative in Milwaukee, Wisconsin. The major investigators, including me, represent psychology, economics, education, anthropology, and evaluation research. We used both quantitative survey techniques and qualitative ethnographic methods.

The New Hope Project was a community-initiated demonstration of an antipoverty program designed to enable full-time workers to move out of poverty. If participants worked full time (more than thirty hours per week), they were eligible for wage supplements designed to raise total family income above the poverty threshold, health insurance subsidies, and child care subsidies. Applicants were randomly assigned to a program group or to a control group.

Experiments always raise ethical issues. The organizers of the New Hope intervention had sufficient funds for about six hundred participants. Rather than simply doing a demonstration on a limited number of people, they chose to conduct a strong experimental evaluation using random assignment of twelve hundred people—half to the program group and half to a control group who continued to be eligible for all other community programs. If the program was successful, there would be a higher probability that it would be adopted widely than would be the case with a simple demonstration. The design was ethically acceptable because the number of participants would be limited; the control group members were not deprived of any benefits that they might otherwise have received; and the design increased the likelihood that the program would be extended. In fact, Wisconsin Works, the new welfare program in Wisconsin, has adopted some features of New Hope.

We did a survey two years after random assignment in which the measures of economic outcomes were accompanied by extensive measures of parent psychological well-being and parenting. We also measured achievement, psychological well-being, and social behavior for children of ages three through twelve. There were large differences between program and control group children on school performance and social behavior (both positive and negative) as reported by teachers. The positive impacts of the program occurred primarily for boys; that is, program family boys performed considerably better than did boys in control group families (Huston and others, 2001). We have administered another survey five years after random assignment as well as doing intensive ethnographic work with a subsample of program and control families.

This research gained the attention of policymakers at the federal level and in many states as well because we formulated the research questions to generate answers that policymakers care about. We consulted with the advisory boards to New Hope throughout the study, putting their issues into a theoretical framework. Our flow chart showing proposed paths of influence on children became a shared joke at advisory board meetings because it was so useful to us and so foreign to them, but it helped to integrate our various perspectives. The advisory boards included community representatives and top scholars in policy analysis. The process of consultation was continuous, from selection of measures to interpretation of the findings. We revised the report in response to reviews by scholars and policymakers so that by the time the results were released, we had developed a receptive policy audience who would consider the findings credible. This process of review and consultation with policymakers is fairly standard in the best evaluation research firms such as MDRC, and we can all learn from it.

This same group of collaborators is now undertaking the Next Generation project, in which we are synthesizing the effects on children and family life from several different large-scale experimental studies of welfare reform and antipoverty initiatives (Morris and others, 2000). Here we have gone one step further to involve policymakers in forming the research questions by asking them what questions they would like us to try to answer. This process functions to make the research relevant to their concerns and create an appetite for the results when they appear.

Designing Research to Inform Policy

Doing research that can influence policy has hazards as well as rewards. You can do harm if your designs are weak or your measures are poor. It can be at least as dangerous to fail to demonstrate a phenomenon that *does* exist—a problem referred to as "Type II error"—as to produce results that appear to demonstrate something that really isn't there (Type I error). A classic example is the Westinghouse evaluation of Head Start, which concluded that the beneficial effects washed out by about third grade (Cicirelli, 1969). These findings led many to conclude that Head Start had no lasting effects; the program was saved from elimination only because there were strong researchers in the federal government (Zigler, Chapter One, this volume) and outside advocates who fought for its continuation.

There is good reason to believe that this study was subject to Type II error. First, all programs called Head Start were combined with little or no information about their duration or quality. Second, later analyses demonstrated that children in *nonrandom* control groups were on average less deprived than were those in Head Start; thus use of these groups underestimated the effects of Head Start (Lee, Brooks-Gunn, and Schnur, 1988). As a result of these design flaws, the early research on Head Start did real harm, and no good.

The best protection against Type II error is use of strong designs, such as random assignment. One of the joys of the New Hope and other Next Generation experiments is the ability to conclude that the treatment *caused* the differences between the experimental and control groups. When random assignment is impossible, quasi-experimental designs and other substitutes for true experiments may be appropriate. It is dangerous to use ad hoc control groups because they may differ from the treatment group in unknown and important ways.

Second, we can minimize Type II errors by using measures with adequate reliability and validity. Specifically, policy-oriented research should use instruments with demonstrated reliability and validity for the population of interest (Busch-Rossnagel, Chapter Five, this volume). Focus groups, pilot testing, and qualitative observations can help interpret the meaning of the chosen measures for the people being studied. One of the major needs in our field is more systematic development of measures, especially for diverse population groups.

Third, our samples should be large and representative of the specified population. Sample size is important because small samples reduce power and raise the probability of Type II error. I often review studies that conclude that their independent variable does not affect the dependent variable when they find nonsignificant differences with thirty or forty cases. If differences are significant with small numbers, they are persuasive, but if they are not, one must be seriously concerned about low power.

For all these reasons, we should routinely do power estimates as well as tests of significance. Power estimates calculated before undertaking a study will tell you how large your sample needs to be to demonstrate an effect, or, if your sample size is fixed, how large the real difference needs to be in order to detect it. For example, in the New Hope study, we needed a sample of 720 to have an 80 percent probability of detecting an experimental-control difference of .18 standard deviations at an alpha of .10. Imagine how large the real difference would have to be with a sample of forty or fifty!

Fourth, we should present findings in a comprehensible metric. It is easier to visualize the meaning of an effect size than an alpha of .05. If the dependent variable has a scale that people generally understand, one can present findings using that scale. For example, my colleagues and I obtained the high school grades of students whose television viewing we had studied as preschoolers. We presented the relations of early "Sesame Street" viewing to achievement in a graph showing the average grade point average (on a scale of 4.0) for adolescents who were in four quartiles of viewing (Anderson and others, 2001). Grades were higher for each higher quartile of viewing time.

Conclusion

Many of us aspire to influence public policy for the welfare of children. Many paths lead toward this goal, and none of them is sure to work. You can raise the odds by choosing questions and issues that have potential for policy applications, by collaborating with people from a range of disciplines, and by doing high-quality work. There is more serendipity in life than most of us like to admit, so be alert to opportunities. Be prepared, too, for setbacks.

With influence comes responsibility. Good research can provide solid information for policy, but poorly designed research can do active harm, especially by producing erroneous negative results. I conclude with the advice given many years ago to my fellow graduate students and me by professor Paul Meehl at the University of Minnesota, "Be soft-hearted, not soft-headed." Applied researchers have a particular obligation to do research meeting the most exacting standards in our field.

References

Anderson, D. R., and others. "Early Childhood Television Viewing and Adolescent Behavior." *Monographs of the Society for Research in Child Development,* 2001, *66*(1). Serial no. 264.

Bandura, A. "Social-Learning Theory of Identificatory Processes." In D. A. Goslin (ed.), *Handbook of Socialization Theory and Research.* Chicago: Rand McNally, 1969.

Bandura, A., and Huston, A. C. "Identification as a Process of Incidental Learning." *Journal of Abnormal and Social Psychology,* 1961, *63,* 311–318.

Bogatz, G. A., and Ball, S. J. *The Impact of "Sesame Street" on Children's First School Experiences.* Princeton, N.J.: Educational Testing Service, 1972.

Cicirelli, V. G. *The Impact of Head Start: An Evaluation of the Effects of Head Start on Children's Cognitive and Affective Development.* Washington, D.C.: Westinghouse Learning Corporation, 1969.

Friedrich, L. K., and Stein, A. H. "Aggressive and Prosocial Television Programs and the Natural Behavior of Preschool Children." *Monographs of the Society for Research in Child Development,* 1973, *38*(4), 1–64. Serial no. 151.

Greenberg, D., and Wiseman, M. "What Did the OBRA Demonstrations Do?" In C. F. Manski and I. Garfinkel (eds.), *Evaluating Welfare and Training Programs.* Cambridge, Mass.: Harvard University Press, 1992.

Gueron, J. M., and Pauly, E. *From Welfare to Work.* New York: Russell Sage Foundation, 1991.

Huston, A. C. (ed.). *Children in Poverty: Child Development and Public Policy.* New York: Cambridge University Press, 1991.

Huston, A. C., and Wright, J. C. "Television and Socialization of Young Children." In T. MacBeth (ed.), *Tuning in to Young Viewers.* Thousand Oaks, Calif.: Sage, 1996.

Huston, A. C., and Wright, J. C. "Television and the Informational and Educational Needs of Children." *Annals of the American Academy of Political and Social Science,* 1998, *557,* 9–22.

Huston, A. C., and others. "Work-Based Anti-Poverty Programs for Parents Can Enhance the School Performance and Social Behavior of Children." *Child Development,* 2001, *72,* 318–336.

Huston-Stein, A., and Higgins-Trenk, A. "Development of Females from Childhood Through Adulthood: Career and Feminine Role Orientations." *Life Span Development and Behavior,* 1978, *1,* 257–296.

Johnston, J., Brzezinski, E., and Anderman, E. *Taking the Measure of Channel One: A Three-Year Perspective.* Ann Arbor: Institute for Social Research, University of Michigan, 1994.

Kunkel, D. "Policy Battles over Defining Children's Educational Television." *Annals of the American Academy of Political and Social Science,* 1998, *557,* 39–53.

Lee, V. E., Brooks-Gunn, J., and Schnur, E. "Does Head Start Work? A One-Year Follow-Up Comparison of Disadvantaged Children Attending Head Start, No Preschool, and Other Preschool Programs." *Developmental Psychology,* 1988, *24,* 210–222.

Liebert, R. M., Sprafkin, J., and Davidson, E. *The Early Window: Effects of Television on Children and Youth.* (2nd ed.) New York: Pergamon Press, 1982.

Morris, P., and others. *How Welfare and Work Policies Affect Children: A Synthesis of Research.* New York: Manpower Demonstration Research Corporation, 2000.

Salkind, N. J., and Haskins, R. "Negative Income Tax: The Impact on Children from Low-Income Families." *Journal of Family Issues,* 1982, *3*(2), 165–180.

Stein, A. H., and Bailey, M. M. "Socialization of Achievement Orientation in Females." *Psychological Bulletin,* 1973, *80,* 345–366.

Stein, A. H., and Friedrich, L. K. "The Effects of Television Content on Young Children's Behavior." In A. D. Pick (ed.), *Minnesota Symposia on Child Psychology.* Vol. 9. Minneapolis: University of Minnesota Press, 1975.

Surgeon-General's Scientific Advisory Committee on Television and Social Behavior. *Television and Growing Up: The Impact of Televised Violence.* Washington, D.C.: U.S. Government Printing Office, 1972.

Wright, J. C., and Huston, A. C. *Effects of Educational TV Viewing of Lower Income Preschoolers on Academic Skills, School Readiness, and School Adjustment One to Three*

Years Later. Lawrence, Kans.: Center for Research on the Influences of Television on Children, 1995.
Wright, J. C., and others. "The Relations of Early Television Viewing to School Readiness and Vocabulary of Children from Low Income Families: The Early Window Project." *Child Development*, 2001, *72*(5), 1347–1366.
Zaslow, M. J., Moore, K. A., Morrison, D. R., and Coiro, M. J. "The Family Support Act and Children: Potential Pathways of Influence." *Children and Youth Services Review*, 1995, *17*, 231–249.

ALETHA C. HUSTON is the Priscilla Pond Flawn Regents Professor of Child Development at the University of Texas at Austin.

4

Doing research in the context of programs designed to promote the psychosocial competence of children and adolescents is an important road to understanding the fundamental anatomy of social development.

Risk and Prevention: Building Bridges Between Theory and Practice

Robert L. Selman

As a developmental psychologist whose research and practice continually inform my theoretical ideas, I am pleased to have this opportunity to share a brief history of my professional career. For over thirty years, our working group (the Group for the Study of Interpersonal Development—or GSID) has conducted research and implemented practice that fosters social competence in children and youth. We have always included a focus on the challenges youth face, be it psychopathology in their families, poverty in their neighborhoods, or prejudice in their society. Here, I will focus on the theoretical work, research, practice, and institutional inventions we have employed to prevent the range of problems that might occur if children grow up without the protection social competence provides.

This chapter will give you a brief yet comprehensive history of the theoretical and practice changes in our model of psychosocial development. The title of this chapter points to my two goals. First, I hope to demonstrate how the road between theory and practice has been a two-way street for me, and how a "research attitude" facilitated this two-way traffic. Second, I will describe how Risk and Prevention (R & P), a master's level program we designed and initiated at the Harvard Graduate School of Education in 1992, evolved to serve as a vehicle to traverse that road. I served as its director through the end of 1999. In addition, in outlining the evolution of our model, the incorporation of theory and practice, not only from my work but also from the work of my mentors and students will highlight the relationship between this research attitude and the creation of R & P.

Two Intellectual and Institutional Mentors of Our Work

My initial interests were documented in my dissertation, on the development of children's social perspective-taking skills and their relation to Kohlberg's stages of moral judgment. After completing my doctoral studies at Boston University in practice-oriented psychology in 1969 I began several years of postdoctoral research with Lawrence Kohlberg at the Laboratory of Human Development in the Harvard Graduate School of Education to focus on child development research. The National Institute of Mental Health postdoctoral fellowship program supported my opportunity to obtain that training. My plan was to work with Kohlberg to elaborate on my conception of perspective taking as defined in my dissertation research. I then intended to bring back to the practice of clinical child psychology a stronger foundation in developmental theory and empirical research.

Lawrence Kohlberg was my first mentor, whose research and dedication assisted in developing the theoretical and practical foundation of the model. For his dissertation, completed in 1958, Kohlberg took on the revision of the work Piaget reported in his classic 1935 monograph, *The Moral Judgment of the Child* (Piaget, [1935] 1965). Using longitudinal evidence from adolescents' responses to hypothetical moral dilemmas, he developed a six-stage theory of moral reasoning. Kohlberg continued to work on the expansion and refinement of his own theory for thirty years.

Our work in its initial phase drew heavily on the strengths of Kohlberg's seminal thinking and research about moral psychology from a cognitive developmental view. In its next two phases, we redesigned our model to address the vulnerabilities of the kind of approach he espoused: the connection of social thought and action, and the relation of developmental and cultural variations in moral values. By the mid-1970s, Kohlberg had already begun to develop a set of guidelines for pedagogical practice. In fact, he had moved in 1968 from the Committee on Human Development at the University of Chicago to the Graduate School of Education at Harvard to test what practical implications his theory had for moral education. Two important strengths of this theory provided the basis of our psychosocial model. A key assumption of Kohlberg's ideas was the importance of perspective taking—he called it role-taking—as a critical psychological, or social-cognitive, skill for the growth of moral reasoning. A second was the importance of positive peer relationships as an essential environmental or social condition for promoting this growth. We began to build our own framework upon these two conceptual strengths.

In 1972, my postdoctoral work was nearing completion, and Kohlberg suggested I see Julius B. Richmond, who was director of the Judge Baker Guidance Center, which had an international reputation for training mental health professionals, and who later was a mentor to many contributors to this volume. I went to see if he might have a position for me. A pediatrician by profession—and previously a medical school dean—Richmond had served as the first director of Head Start. He had also served as president of the Society for Research in Child Development (in 1968). In addition, Richmond chaired the Child Psychiatry Service at Boston Children's Hospital, across the street from the affiliated Judge Baker, and was a professor at the Harvard Medical School just down the block.

Richmond hired me as a staff clinical psychologist half-time at the Judge Baker Center. There, I was able to take the theoretical and research skills I had learned working with Kohlberg and put them to use in clinical practice, working with troubled children. My duties were that of most junior staff clinicians: to provide psychological treatment for children, either in the Judge Baker outpatient department or at its on-site clinic school for children with social and emotional problems, the Manville School. However, it was not my earlier predoctoral professional psychology training that motivated Richmond to hire me, it was my postdoctoral research experience. He believed the Judge Baker, with its close affiliation with a major research and teaching hospital for children, ought to be doing more than providing service for children. It ought to be doing research in their service as well. This encouragement began the journey of developing the psychosocial model.

Although it fell within the sphere of influence of the university and was literally in the shadow of the Harvard Medical School, the Judge Baker was still primarily an independent social service agency without resources to support research directly. In addition to my clinical responsibilities, I would have to find the time—and the money—to do the research expected of me on my own. Kohlberg, generous to a fault, supported the other half of my time as a research associate for another year—with very few responsibilities in his lab—while I wrote proposals to support our research, and my salary. Fortunately, with a grant from the newly founded Spencer Foundation, in 1973 we began in earnest to study the growth of interpersonal understanding in both well-adjusted and troubled children, ages six to sixteen. We founded the GSID, located it between the two institutions, and began to interview children of different ages and clinical status about their views on social relationships such as friendships. I found in Richmond a second mentor.

Phase One: Research on the Growth of Interpersonal Understanding

From the start our work maintained its connection to practice. Our practice was of two kinds. One was clinical, an attempt to use developmental theory to understand—and modify—the interpersonal understanding of the troubled children we were seeing in individual therapy. The other was educational, to promote children's social and moral reasoning through classroom discussions of Kohlberg-like social and moral dilemmas designed for children in elementary and middle school. The opportunity for applied developmental work in the child clinical setting was obvious. Yet it was less obvious what kind of practice-based research to do in schools.

Fortunately, in late 1970, halfway through my postdoc, an educational publishing company that developed audiovisual materials for classroom use approached Kohlberg to design some moral dilemmas for the elementary grades. Since Kohlberg's primary interest at the time was on the shift in moral awareness from the middle adolescent period toward young adulthood, the construction of social and moral dilemmas both challenging and appropriate to children in the elementary grades fell to me.

A good interpersonal or moral dilemma provides difficult choices for action. It should put the participant in some conceptual conflict, and it should generate disagreement among the members of a specified population (for example, children aged five to seven). I decided to focus on interpersonal dilemmas where the coordination or balancing of social perspective was key. For example, in one dilemma we constructed, Kathy must decide whether to go on a previously planned play date with a close old friend or to a special event with a new one. This dilemma was not predominantly moral, but it definitely was peer oriented. It involved Kathy considering the three points of view (self, old friend, and new friend) in the process of making a choice. The publishing company marketed the audiovisual depictions of the dilemmas as "practice"—that is, as supplemental social studies curricula. We used them as part of our interview research protocol with children at different ages to launch the research work on the growth of interpersonal understanding.

Our task was to compare the interpersonal understanding of "well" socially adjusted and "less well" socially adjusted children. We drew the sample of less well-adjusted children from the day treatment school in the Judge Baker, where now I was on staff. This was how one did clinical research in the framework of an academic medical model. The researcher could take advantage—in the better sense of the term—of the accessibility of the patients and clients in the clinic or hospital as subjects and participants. In theoretical terms, we also called the work clinical-developmental. Here, it meant something a bit different: using a developmental framework in the comparative study of the social competence of children growing up.

By the close of our first phase of research, our theoretical framework was a clinical-developmental design. We used the framework to compare the social awareness of troubled kids over time with that of their peers. We constructed a scoring manual of more than three hundred pages that gave examples of children's conceptions of interpersonal relationships—that is, of friendship, peer group, and parent-child relationships at each level of perspective coordination in the developmental framework.

We used our core theoretical construct, the capacity to coordinate social perspectives, the practice-based dilemmas, and our manual to assess the interpersonal understanding of children at different ages. What proved surprising and led to our next phase, as well as strengthened our commitment to linking theory and practice, was the unexpected cases in which high levels of interpersonal understanding did not map directly to adjusted peer relations.

Our research indicated that there were some children whose understanding of relationships in the social world could be perfectly normal—or even advanced for children their age by our developmental standards—who were socially reclusive, lonely, and isolated. Still others with age-appropriate interpersonal understanding could be feared, hated, and despised because they were bullies, or impulsive and aggressive. What could account for the gap between their competence level of social thought and their quite variable or oscillating level of social action?

Phase Two: How the Practice of Pair Therapy Helped Us Construct a Developmental Model of Social Interactions

Several months before he enlisted for another tour of public service—this time as Surgeon General and Assistant Secretary of Health in the Carter Administration—Richmond asked me to direct the Manville School of the Judge Baker. In a high-powered, research-oriented medical school model, he told me, the director of the hospital service unit is responsible not only for the clinical service (patient care) and clinical training (teaching and supervision) of young professionals, but also for starting and fostering research initiatives relevant to the kinds of problems the service treats. Richmond saw no reason why this model should not apply to schools as well as to hospitals, to research in education as well as in medicine. Here I saw the opportunity to combine the work on our model with clinical treatment.

Often the emphasis in the medical model of clinical research is to find causes and cures for the diseases that the hospital service treats. At the Manville School, we were treating a large number of children with psychological ailments that affected their social relationships. This fell under the discipline of child psychiatry within the medical model. Mostly we were treating these student patients with either individual psychotherapy or family therapy heavily oriented toward parent-child relationships. Almost without exception, however, children referred to our school had great difficulty getting along with peers. The treatments we had put in place did not focus on how to help children get along with peers at a time in life when peer relationships often are primary. The challenges became how to directly address problems involving the disorders these children had in the capacity for maintaining both intimacy and autonomy in close peer relationships—that is, for making and keeping friends.

In 1979, as we were completing the summary of our descriptive and comparative clinical-developmental research on the growth of interpersonal understanding, we decided to attempt to integrate what we had learned about theory and research with practice. This time, the practice would take the lead. My role as director of the Judge Baker's Manville School facilitated our capacity to develop psychosocial treatments targeted directly at children's difficulties with peer relationships. Our aim was not only to raise their level of social understanding but also to close the gap between their improved social thought and their social interactions, especially with peers. The question became, What kind of treatment would help promote higher levels of social understanding and action in those children who lacked either or both?

Thus the associated theoretical question became how to conceptualize, within a developmental framework, the relationship between thought and action in the social domain. Could we measure levels of social interaction skills and performance in ways that were theoretically compatible and empirically comparable with the assessment of levels of social thought we had defined in the previous phase of research? If so, perhaps we could begin to address one of the two main problems Kohlberg's opponents had with his cognitive framework—how to study its link to social action. With a framework that could compare levels of thought and action, we could also study the gap between them, rather than viewing the gap as an invalidation of the framework itself.

Our primary approach to the treatment of children and adolescents with serious difficulty making friends we have written about extensively elsewhere (Selman and Schultz, 1990; Selman, Watts, and Schultz, 1997). We called it *pair therapy*. Through our systematic research and observation of social interactions in the practice of pairs, we began to conceptualize the intimacy and autonomy functions of dyadic social relationships between and among peers. We generated a developmental language to describe attempts on the part of the partners in the pairs to keep a psychological distance from each other or to become closer. We called the former "strategies for interpersonal negotiation," and the latter "forms of shared experience." In this sense, the levels and orientations of social action strategies that emerged out of the practice of pair therapy provide a clear example of how practice-based research can transform theory.

Clinical practice informed our model, actually articulating two dimensions in two ways. First, in the new domain of the skills to manage interpersonal relationships, it developmentally defined (that is, by level) the complementary nature of intimacy and autonomy functions in friendships at each level of perspective coordination. Second, within each function, it integrated a developmental source of variation in social competence—level of social perspective coordination—with another variation in social behavior that often is even more visible to the naked eye, especially in the observation of pairs. We called this dimension "interpersonal orientation," meaning whether or not the social action an individual takes—in social relationships with "the other"—is oriented to accommodation to the other or to transforming the other to accommodate to the self. (For additional information, see Selman and Schultz, 1990).

This integration of the interpersonal orientation dimension with our earlier description of developmental levels was crucial to our new and revised formulation of how to study thought and action within one framework. Actions that looked quite different from one another, for example, flight or fight, could nevertheless be at the same level, developmentally speaking, for example, impulsive and unilateral. The opposite was also possible: similar-looking actions, say, leaving a counseling room because of a conflict, could be driven by developmentally different reasons, for example, impulsive or strategic. At last we had a conceptual tool designed to tell the difference.

Phase Three: From Treatment to Prevention Through an Institutional Intervention

In the mid-1980s, the Judge Baker shifted its primary focus to innovative, research-based intervention and psychosocial prevention projects. It even changed its name to the Judge Baker Children's Center to emphasize its work to help all children, but especially those growing up in low-income communities, meet the challenges of urban life. This shift was based on the observation that most of the children's diagnosed psychiatric disorders were highly preventable if the causes could be detected early and effective psychosocial interventions were available to reduce the risk factors. Thus, we expanded our service focus from the psychological treatment (in our clinic special needs school) of severe psychiatric disorders to the psychosocial prevention of these disorders in public schools and community-based mental health services.

The attempted institutional move—from treatment in the clinic to prevention in the community—required a theoretical move as well. We needed to figure out a better way to bring an understanding of the cultural backgrounds of the children we worked with into our evolving developmental framework. We needed protection against interpreting cultural variations in children's social awareness as sick or immature.

By the mid-1980s, I divided my time equally between research and graduate teaching at Harvard and practice at the Judge Baker. This new balance facilitated a three-way institutional partnership that trained a small group of doctoral students in the treatment—and the prevention—of child and adolescent psychopathology. The Judge Baker was the lead agency in this partnership. The Counseling doctoral program in the Harvard Graduate School of Education was one partner. The Boston Public Schools (BPS) was the other partner. This three-way partnership model—university (training of professionals), mental health agency (supervision), and public school (point of service)—evolved into the model for partnerships within the current Risk & Prevention master's program.

Much of the impetus for this tripartite model of psychological training came from the graduate students themselves. In the late 1980s, doctoral students knew that children growing up in poor, urban, minority communities faced risky circumstances far beyond family dynamics. They were coming for their doctoral training already highly committed to providing accessible services for this underserved population. In many ways, they were the prototype of the kind of master's level students who would apply to Risk & Prevention in the 1990s. They did not want to think of the kids they were working with as growing up "disordered." They wanted to think of them as "placed at risk," as needing some support growing up in disordered communities or in a society that discriminates, in the negative sense of the term, against them.

Advanced doctoral students first worked in "tertiary prevention" at the Judge Baker Manville School for a year as psychology interns, providing psychological treatment for children with severe psychiatric disorders. Pair therapy was part of this regimen. Then the following year they provided psychological services for targeted children at high risk for developing psychosocial problems (secondary prevention) at selected Boston public schools. They also continued as interns at the Judge Baker, where senior staff clinicians, supported through public funds and private foundation grants, provided supervision for services provided at both clinic and public school sites. What these doctoral students learned at the Judge Baker as a treatment center for the cure of psychiatric disorders, they adapted and applied to the public schools as institutions for prevention of these disorders.

The lessons learned about the risks of students in urban public schools by the doctoral students who ventured to and through the Judge Baker for their training played a major role in the subsequent design process for Risk & Prevention master's level program. Through carefully designed community based experiences, the program integrates research and practice, quantitative and qualitative methods of analysis, and perspectives on practice from different social science disciplines such as cultural anthropology and developmental psychology. Each of its prevention approaches orients toward the integrative promotion of children's academic and social competence.

The R & P specialization replicates in a range of settings—from infant and toddler centers for children at risk to elementary, middle, and high schools—the tripartite partnership model. University, community-based mental health agency, and public school worked together as partners. Each partner brought its own interests and expertise to the promotion of social and academic success for children and adolescents placed at risk. Faculty, research-oriented doctoral students, and practice-oriented masters' students would all share a common purpose, helping kids placed at risk succeed academically and socially. Although this program is much broader than our evolving research-theory-practice framework, our own research agenda would find this a meaningful place to operate, grow, and develop alongside those of our colleagues.

In the 1990s, I shifted most of my time to the Harvard Graduate School of Education. I wanted to focus more on prevention, and I felt education was a good place to do it. In 1992, after two years of pilot work, we founded Risk & Prevention. Each of its projects, which is sponsored by a member of the faculty, bridges theory and practice and provides an orientation to both developmental and cultural influences on, and analyses of, the meaning children and adolescents make of the risks they face growing up. Projects spawned by the R & P faculty include the Developmental Pathways Project of Cathy Ayoub and her colleagues (1996), Project IF, initiated by Michael Nakkula and S. M. Ravitch (1997), and The Rally Project, started by Gil Noam and his colleagues (1996).

To help children and adolescents, each project also designs psychosocial prevention practices that are "meaning oriented." Rather than only focusing on the prevention of psychosocial problems through providing youth with factual knowledge of possible negative consequences of taking risks, these approaches begin by seeking to understand what the risky behavior means to children. The same risk behaviors can mean different things to children at different ages, in different contexts, and in response to different backgrounds.

From Practice Back to Theory and Research in Phase Three

So far, I have tried to emphasize the reciprocity between the institutional mission of R & P as an applied developmental psychology program at the graduate level and the focus of the research done within it. Earlier, I expressed the opinion that research on social development out of context is not usually very meaningful or relevant. We have pushed this idea a bit further to argue that research on social development, broadly defined, thrives in a context where connections are made to practitioners focused on fostering it. What then was the influence of the move to prevention practice in the community on our own theoretical framework during the third phase?

Practice journals and reflective papers from R & P students are striking examples of data that point to the importance of a framework within which one can listen to children and adolescents speak about the risks they face in the context of their interactions and relationships. For example, after guiding pair counseling (see Selman, Watts, and Schultz, 1997, for how we moved from work on pair therapy in treatment settings to pair counseling in school-based prevention) over the course of a year, a graduate student, Kesia Constantine, offered an analysis of the cultural and personal meaning of fighting for her clients:

> Mark was referred to pair counseling (with Lance) for what would be classified as conduct disorder behavior. They are both fourth-grade males ten years of age, both of whom are actively engaged in physical fighting. Mark is a

> highly regarded fighter among his classmates; it targets him as an individual to be fearful of, to challenge, or to look up to. At school, at least among the boys, physical fighting is a means by which one gains and loses status.
>
> It has been a difficult process to get Mark to reflect directly on his engagement in physical fights. My information has come through other avenues. Through Mark's teacher, for example, I learned Mark feels very unsafe in his neighborhood. Some older boys frighten Mark. He would walk several blocks out of his way to avoid them, or even stay home. Yet he is quite concerned not to appear to be a "punk" in the eyes of others in his community.
>
> Mark refuses to back down from any challenge, even when falsely accused. I was beginning to be frustrated by episodes of frequent fighting on the bus, because it seemed as though Mark, Lance, and the other boys know what it takes to stay out of trouble on the bus, yet they actively went out of their way to find trouble. The teacher viewed Mark as the instigator of many of the fights and seemed to think that he fights because he knows he can win. In contrast, Mark sees things quite differently. He sees himself as getting into fights only when he is challenged and he cannot back down; he doesn't want to look like a wimp.
>
> It became apparent to me in our conversations about fighting that the boys get the same rush from fighting as they do from playing sports. For the boys, sports represent a chance to be powerful, the ability to prove that they are better than someone else by winning the game. Skill speaks for itself, so they don't need to fight. Over the course of my year with Mark and Lance I realized that fighting shouldn't simply be thought of as a risk-taking behavior that they engage in, but rather it should also be thought of as a skill, something to take pride in.

This excerpt demonstrates what pushed us to listen to youth for something beyond their theories of interpersonal understanding and their repertoires of social strategies developed in pair therapy. It pushed us to direct our developmental analysis toward the personal meaning—always culturally derived—they make of the actual risks they face every day in the context of their own personal social relationships.

Although the sources of influence on our theoretical framework are multiple, the effects on it in this third phase are clear. We became more aware of the role the personal meaning of their relationships plays for children and adolescents in the risks they take. We began to use data observed to developmentally analyze the moment-to-moment levels of awareness children have of the risks they take in connection to their most important social relationships. For instance, the Kesia Constantine's summary of progress over the year for Mark concluded this way:

> Mark has different reasons than before for not fighting. Previously, it was only when "no one was messing with me." Now the meaning of fighting has changed, not just conceptually, but personally within him. He says he does

not fight because "he does not need to fight to prove he is cool." He is showing greater self-awareness.

We have incorporated this kind of practice-based evidence into our theoretical analysis. We have integrated a developmental analysis of the personal meaning of risky behavior, such as fighting, with a cultural one. Mark's self-awareness has become what we call "need based," rather than simply "rule based." By this, we mean his actions are no longer simply based on a level of awareness limited to vague perceptions of the force of a cultural rule like "wimps don't fight." They were becoming more often—but certainly not always—based on his greater reflective understanding of his own personal needs. New self-awareness gave him more control of his own actions and more insight into the risks of fighting. In pair counseling he learned that fighting could mean losing a friend, while not fighting was still the greater risk in the neighborhood because it meant loss of power and status in the peer group. Our research question now is how powerful is this evolving self-awareness as a form of protection against the slings and arrows of everyday existence.

Summing Up

Here in the institutional context of Risk & Prevention, our practice-based research again drove the revision of our theoretical framework. We now analyze three dimensions of the growing psychosocial development of children: levels of social understanding, repertoires of social skills, and the awareness of the personal meaning of risks—within social relationships (Levitt and Selman, 1996; Selman and Adalbjarnardottir, 2000). This generates our current conical, three-dimensional framework for analysis, as portrayed in Figure 4.1. This revised model, we hope, also will point the way to meaning-centered interventions to help children develop the depth of self-awareness to say what they mean about the risks they take—and mean what they say. We believe such awareness is good prevention.

We also believe it is important to embed research on children's negotiations in their social world in practical contexts that promote engagement. In so doing, research gains relevance and enhances the opportunity to provide additional validity for the theory to which it is connected. It increases the likelihood that the theory will be more enduring because it inoculates it against the faddishness of research styles and theoretical fashions that travel only in closed academic circles. If theory and practice are paired in this way, practice-based research helps to bond them. This has been the primary focus of our work over thirty years. Theoretical changes in our model attempt to maintain a developmental framework in light of the challenges of linking thought and action and connecting cultural and developmental variations in each. However, with each major design shift in our model, understanding the developing capacity to coordinate social perspectives, which is where I started my work, is still at its core.

Figure 4.1. A Developmental View of Risk (for example, "Mark or Lance fighting") and Social Relationships (for example, friendships and peers): The Analysis of Three Psychosocial Competencies

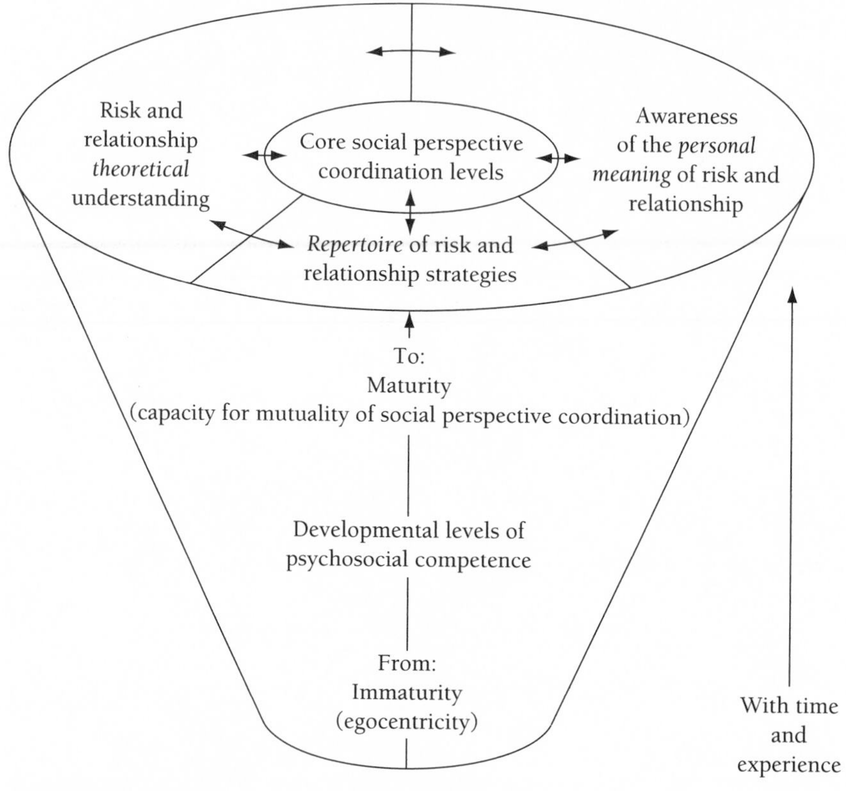

Source: Selman and Adalbjarnardottir, 2000, p. 54. Reprinted by permission of Lawrence Erlbaum Associates, Mahwah, New Jersey.

References

Ayoub, C., Raya, P., Miller, P., and Geismar, K. "Pair Play: A Relational Intervention System for Building Friendships in Young Children." *Journal of Child and Youth Care Work,* 1996, *11,* 105–118.

Levitt, M. Z., and Selman, R. L. "The Personal Meaning of Risky Behavior: A Developmental Perspective on Friendship and Fighting." In K. Fischer and G. Noam (eds.), *Development and Vulnerability*. Hillsdale, N.J.: Erlbaum, 1996.

Nakkula, M. J., and Ravitch, S. M. *Matters of Interpretation: Reciprocal Transformation in Therapeutic and Developmental Relationships with Youth.* San Francisco: Jossey-Bass, 1997.

Noam, G., Winner, K., Rhein, A., and Molad, B. "The Harvard RALLY Program and the Prevention Practitioner: Comprehensive, School-Based Intervention to Support Resiliency in At-Risk Adolescents." *Journal of Child and Youth Care Work,* 1996, *11,* 32–48.

Piaget, J. *The Moral Judgment of the Child.* (M. Gabain, trans.). New York: Free Press, 1965. (Originally published 1935.)

Selman, R. L., and Adalbjarnardottir, S. A. "A Developmental Method to Analyze the Personal Meaning Adolescents Make of Risk and Relationship: The Case of Drinking." *Journal of Applied Developmental Science,* 2000, 4(1), 47–65.

Selman, R. L., and Schultz, L. H. *Making a Friend in Youth: Developmental Theory and Pair Therapy.* Chicago: University of Chicago Press, 1990.

Selman, R. L., Watts, C. L., and Schultz, L. H. (eds.). *Fostering Friendship: Pair Therapy for Treatment and Prevention.* Hawthorne, N.Y.: Aldine de Gruyter, 1997.

ROBERT L. SELMAN is Roy Edward Larsen Professor of Education and Human Development and professor of psychology (psychiatry) at Harvard University. He currently serves as chair of the Human Development and Psychology Area in the Harvard Graduate School of Education.

5

Research with ethnic minority and subcultural groups is often stymied by measures developed on samples of European Americans. This chapter presents recommendations to help researchers create measures that are sensitive to both the culture and the communities of research participants.

Creating Culturally Sensitive and Community-Sensitive Measures of Development

Nancy A. Busch-Rossnagel

In 1993, I received a grant from the National Institute of Child Health and Human Development to study socioemotional development in Dominican and Puerto Rican toddlers. I was particularly interested in how the socializing environment provided by these children's families would influence the development of mastery motivation and related aspects of the self-concept (Busch-Rossnagel, Knauf-Jensen, and DesRosiers, 1995). However, my ability to explore such influence was limited by the dearth of instruments to assess several of the constructs of interest, such as child-rearing practices, that would be part of the socializing environment in Puerto Rican and Dominican families.

Many colleagues are amazed when they look carefully at the goal of the grant and realize that I was funded to create culturally appropriate measures of child-rearing practices and toddler development. Their amazement comes from the conventional wisdom that you can't get funding for instrument development. However, my experience contradicted that conventional wisdom because, I believe, I clearly documented the need for new instruments and articulated the methods for creating valid measures. Many people acknowledge the need for new or revised instruments that reflect the cultural diversity of the population, but they don't know how to address this need. The point of this chapter is to share with you the steps we used to create measures that are culturally appropriate and sensitive to the needs of the communities in which we work. In my lab, we developed both standardized observational techniques and questionnaire measures (DesRosiers and others, 1999; Dichter-Blancher, Busch-Rossnagel, and Knauf-Jensen, 1997). I will be sharing examples of the creation of the questionnaires because they are harder to develop.

Apply Practices from Other Fields

The first suggestion I would make to developmental psychologists is to combine our own best practices with the contributions of other fields in identifying good practices for working with subcultural and community groups. One of the strengths of applied developmental psychology is its rigorous research methodology, particularly that of the classic experiment. Much of the developmental research on subcultural or ethnic minority groups uses this approach to make comparisons between majority samples of European Americans and minority or subcultural groups, such as Puerto Ricans and Dominicans.

However, from the standpoint of experimental design, these comparisons are flawed. Because we cannot randomly assign majority or minority status or cultural background, our research designs do not control potential confounding variables, such as language, education, income level, and so on. Because of these confounds, such comparisons do little to enhance our understanding of the role that psychological processes underlying culture play in development (Busch-Rossnagel, 1992).

This is where another strength of applied developmental psychology, its acquaintance with other disciplines, can play an instrumental role. In my research with subcultural groups, I used a concept from cultural anthropology and cross-cultural or ethnopsychology: contrast cultures. Contrast cultures provide a different way of looking at majority-minority comparisons. Much research done in cultural anthropology is done with cultures that are strikingly different from the American culture. A culture that is significantly different from your own provides an opportunity to examine the ethnocentricity of the assumptions of your research.

In cultural psychology, the dimension that extends from an individualistic to a collectivistic orientation is receiving considerable attention. On this dimension, traditional Hispanic cultures represent a contrast to the predominant European American culture of our society (Triandis, 1989). American culture has been characterized as an individualistic culture, one that emphasizes independence and separation from relationships such as family, community, and clan. In contrast, Hispanic cultures are seen as collectivistic or interdependent, with an emphasis on familial and communal relatedness.

In addition to the notion of cultural orientation, anthropology and cross-cultural psychology are full of examples of *participant observation* or *ethnography*. In ethnography, the researcher lives in the community and gets the insider view of the group. The experiences are documented through careful observation. Ethnography is very expensive in terms of time: observation often takes years. I didn't have the luxury of years to learn about Latino socialization efforts, but I did emphasize careful observation in my research with Puerto Rican families. Such observations were responsible for enlarging our concept of mastery motivation to explore the relationship to social interactions and self-concept.

Mastery motivation is the impetus to achieve and improve one's skills in the absence of any physical reward—the mastery of the environment seems to be the reward in itself (Busch-Rossnagel, 1997). Examples of mastery motivation would be learning to put pieces in a puzzle or finding out how a toy CD player worked. Most research on mastery motivation focused on young children working by themselves to explore and use a toy in a goal-oriented or task-directed manner (Morgan, Busch-Rossnagel, Maslin-Cole, and Harmon, 1992). The standardized method used to measure mastery motivation emphasized the children's independent actions, so we quantified mastery motivation as children's persistence in working on a task and their pleasure while doing so.

Testing Puerto Rican toddlers with our standardized measures of mastery motivation, we carefully observed the reactions of Puerto Rican children and noted that they showed much more social referencing than the Anglo toddlers I had tested before (Busch-Rossnagel, Vargas, Knauf, and Planos, 1993). Social referencing is when a young child looks to a more knowledgeable partner to interpret a situation, as when children who fall look around for a parent's reaction. If the response is a laugh and a smile, then the child laughs. But if the parent says, "Oh no! Are you OK? Don't cry!" then the child will start to cry.

The standardized testing of mastery motivation called for adults to redirect the child's attention to the toy and to encourage task-directed actions with the toys. However, our observations showed us that the Puerto Rican children often persisted in attempting to interact with the adults. I started wondering if the children were more motivated toward interactions with people than toward mastery of objects.

Theoretical work (Harter, 1978) suggested that mastery motivation can be observed in several domains, and several researchers (Barrett and Morgan, 1995; Combs and Wachs, 1993; MacTurk and others, 1985) had expanded their notion of mastery motivation in young children to include actions directed toward people. This expansion of the concept of mastery motivation seemed to fit with our observations of Puerto Rican toddlers. When adults had redirected the toddler's attention back to the toys, some toddlers focused again on the task associated with the toy (fitting the shapes into openings on a shape sorter). However, other toddlers created their own tasks, and most of these tasks involved social interactions. The best example of this was one boy who used the shapes for the shape sorter to create an imaginary zoo. The animals interacted with each other, had to go to sleep, and so on!

The members of my lab discussed these observations of social interactions and eventually proposed the hypothesis that social mastery motivation may be more compatible with collectivistic cultures than object-oriented mastery motivation. Thus, if we were to understand the development of mastery motivation in Dominican and Puerto Rican toddlers we had to look at social mastery motivation as well as object-oriented mastery motivation.

I could have spent years making observations of young children aged fifteen to forty-two months in social interactions to develop a measure of social mastery motivation. To speed up the process, we turned to the experts, Puerto Rican and Dominican parents of young children. We used the methodology of *focus groups*, a technique common in social psychology. This method is ideal for a study that seeks to identify constructs because it avoids experimenter bias by allowing participants to express themselves freely; the facilitator of the focus group keeps the discussion on the topic, but does not lead the discussion.

We used a series of focus group discussions to identify examples of social mastery motivation and to gather information about the socialization environment. In the first set of focus groups, we asked Puerto Rican and Dominican parents to react to some vignettes describing the behaviors of young children in social situations. We also asked them to describe some similar situations with children that they had observed. From our focus groups we identified six domains of socioemotional development related to mastery motivation: persistence directed toward objects, social mastery with peers, social mastery with adults, autonomy or self-assertion, self-evaluation, and self-regulation. For each of these domains, we developed several additional vignettes of child behaviors that exemplify each domain. (For a discussion of the relationship between mastery motivation and self-concept, see DesRosiers and Busch-Rossnagel, 1997.)

These vignettes were then used for the next set of focus groups, where we asked Puerto Rican, Dominican, and Mexican parents to discuss how they would respond to the children's behavior in the vignettes and to identify the appropriate socialization goals in each domain (see Busch-Rossnagel, Vargas, and Knauf-Jensen, 1995). The transcripts of these discussions provided the raw data to develop items for our questionnaires, and the content analysis of the statements in the focus group discussions helped us design the test blueprints for the questionnaire measures.

Keep Psychometric Rigor

A *test blueprint* is just what the term brings to mind, a guide or a blueprint that shows the outline of what the finished test or assessment instrument should look like. The blueprint identifies the constructs to be tapped by the test and the domains in which the construct will be assessed. We applied the technique of classical test theory in several stages of our instrument development for our measures of child-rearing practices. We first used the content analysis of the second set of focus group discussions on parenting behaviors and socialization goals to identify four categories of statements: descriptions of the parents' behaviors, descriptions of what

Table 5.1. Test Blueprint for the Socializing Environment Questionnaire (SEQ)

	Domains of Socioemotional Development					
Parent Response	*Persistence with Objects*	*Social Mastery with Adults*	*Social Mastery with Peers*	*Autonomy*	*Self-Evaluation*	*Self-Regulation*
Directive						
Explanation						
Bargaining						
Threat						
Negative nonverbal intervention						
Punishment						
Positive nonverbal intervention						
Patience						
Compliance						

they had seen other parents do, child behaviors that were desired or not desired as socialization goals, and cultural values that were desired socialization goals (for example, respecting adults) (Busch-Rossnagel, Knauf-Jensen, and DesRosiers, 1995). These types of statements reflected three different constructs related to child-rearing: parenting, child behaviors, and cultural values. We needed three separate measures to assess these different aspects of child socialization.

The statements about parenting were used to develop our Socializing Environment Questionnaire (SEQ), which was designed to assess parent behaviors related to mastery motivation. The test blueprint of the SEQ (presented in Table 5.1) identified two constructs, parent behaviors and mastery motivation, that needed to be sampled adequately to ensure adequate content validity for the measure. The content analysis of the focus groups on parenting behaviors showed that six specific behaviors accounted for half of the parent statements. These were directives, compliance, tolerance, help, explanations, and bargaining. For other-parent behaviors, there were numerous examples of directives, but less positive behaviors, such as physical punishment, negative physical interventions, and refusals, were also mentioned frequently.

The final test blueprint for the SEQ assesses nine parent behaviors within six specific domains of socioemotional development; each domain is tapped by two vignettes. To select the vignettes for inclusion in the final version of the SEQ, we asked graduate students to assess the match of vignettes to each domain. The students were provided with the label and definition of each domain and asked to rate how well each vignette tapped each domain. The index of item-objective congruence (Rovinelli and Hambleton, 1986) was used to quantify this assessment of content validity, and the two vignettes with the highest index were selected to tap each domain. Table 5.2 lists the definitions of each domain and provides one vignette that taps the domain.

Table 5.2. Vignettes Used to Exemplify Domains of Socioemotional Development

Domain and Definition	*Vignette*
Persistence with Objects: Child continues own activity even after requests to stop or offers of assistance.	Just recently, for his birthday, Carlos got a new toy that does several different things if you push the right buttons. Carlos really seems to enjoy playing with this toy although he can only get one or two things to work. When his mother suggests that it is time to go out and play, Carlos wants to keep playing with the toy.
Social Mastery with Adults: Child learns the skills of interacting with adults.	A friend of Ashley's mother comes to visit. She meets Ashley for the first time and gives her a kiss. Ashley, who is two and a half years old, does not look at or talk to the friend. Ashley's mother thinks her daughter should respond, but is unsure about what to do.
Social Mastery with Peers: Child learns the skills of interacting with peers.	Marcos is an only child who loves to watch his three cousins play, and pulls his mother to come closer to the group of playing kids. However, Marcos won't join in the play even though his cousins are about the same age. His parents aren't sure what they should do.
Autonomy: Child does something without assistance or takes the initiative to act independently.	Blanca is a three-year-old who is always doing something and looks for every opportunity to be involved in everything that is going on. For instance, last night, Blanca insisted on bringing the dishes to the table, saying, "I do it." Blanca seemed to enjoy this activity, although her parents had to wait a while longer to sit down to eat.
Self-Evaluation: Child internalizes pride and shame.	Luis is a three-year-old who was given a toy guitar as a present. When he thinks that nobody is watching he sings and plays his guitar. However, as soon as someone walks into the room and says something about his singing, he smiles shyly and stops what he is doing. His parents are concerned about their child being so self-conscious or shy.
Self-Regulation: Child learns to modulate behavior to conform to societal expectations.	Susana, age three and a half, likes to play with dolls. Her parents have bought a very expensive porcelain doll and told Susana that she should not touch it. Left alone for a few minutes, Susana starts to play with the doll. When her parents return, they notice that Susana has been playing with the doll.

Deal Simultaneously with Both Cultures

We often used the parents' own words from the focus group discussions in writing the initial set of parental response items for the SEQ. In many cases, those words were in Spanish, so we had to face the question of whether to create a Spanish questionnaire or an English measure or both. The researcher who really wants to create culturally sensitive measures has to face up to the thorny problem of linguistic equivalence. From my experience this is where many measures fall short because the issue of linguistic equivalence cannot be dealt with adequately after the fact.

The creation of linguistically equivalent measures begins with translation by asking a bilingual individual to prepare a version in the second language. But many words have different nuances: for example, *practice* can mean carry out or rehearse. How do you know whether the translation has the proper meaning? One way is to complete a *back translation* by taking the Spanish version and translating it into English. If the two versions are different, you can alter the Spanish to more closely approximate the original English. Once the Spanish version is altered, another back translation to English is done. An iterative process, going through several rounds of translation and back translation, usually improves the comparability of the Spanish and the English versions (Marín and Marín, 1991; Werner and Campbell, 1970).

Back translation with several iterations is usually seen as the best practice to develop linguistically equivalent versions of measures. However, because only the Spanish version is modified—the English version is not changed—back translation has limitations. When the original English measure is standardized and cannot be modified without jeopardizing the psychometric information gathered, then back translation must suffice.

However, when both versions of the instruments are being developed simultaneously, a better option is available, the process of *decentering* (Werner and Campbell, 1970). On the surface, the process of decentering is the same as the iterative process of back translation. However, either version may be modified to enhance the match between the two. Where discrepancies exist between the two versions, you can discuss the intent of the English item, rewrite the item for clarification, and then translate and back-translate again. In other words, each round of translation informs the development process for both versions of the questionnaire and often has the effect of clarifying the focus of the items.

Decentering affected the development of our items, particularly on the questionnaires. For example, the initial form of one of our social mastery motivation items was, "Dislikes make believe and gives up quickly in playing it with adults." The initial Spanish version was even more awkward, "*No le gustan los juegos de fantasia y se do por vencido cuando juega con adultos.*" The process of decentering produced the following items: "Gives up quickly in pretend play with adults." And "*Se da por vencido rapidamente cuando juega juegos de fantasia con adultos.*" The decentering provided us with a simpler version in both languages. The language was more straightforward and at a lower reading level. Decentering removed awkward wording in Spanish and clearly placed the emphasis on "giving up quickly," which was the point of the item.

In addition to the process of decentering, we also did analysis with bilinguals' completing both the Spanish and English forms to assess the equivalence of the forms. We continue to examine the pattern of relationships for the two versions to see if they are psychologically as well as linguistically equivalent (Knauf-Jensen, Busch-Rossnagel, and Morgan, 1997). As with any type of psychometric effort, linguistic and psychological equivalence are never proven; we just continue to gather more evidence.

Develop Community Partnerships

Once they have a decentered instrument, many researchers feel that they are finished. However, I suggest that researchers still need to consult with the research participants about the effect of the instruments—and the research—on the individuals and the community itself. Perhaps what is most unique about our instrument development efforts is that we go back to community centers to ask the mothers about our efforts. We ask the mothers and staff whether they think that we have captured their behaviors. In effect, we ask for their evaluation of the face validity of our instrument, a type of validity not accepted from the vantage point of psychological science, but clearly an issue when it comes to how individuals respond to research situations.

The distrust of research, which our research team, like many others, initially found, may come from poor experiences in the past. Research participants told us of other researchers who promised to share the results of studies, but never did. I assume that they ran out of time (or money) to do so, a situation that students often face, and too characteristic of research in general. Nonetheless, such feedback is a critical component of culturally sensitive research. Without the feedback, research participants may be left feeling exploited and become even more distrustful of future research endeavors. In contrast, when we take our instruments back to the participants—with their own words in the items—the mothers understand that we take seriously what they tell us. I believe that they begin to understand the role that research can play in answering the questions that they have about raising their children.

This continued collaboration is the heart of my model of culturally sensitive instrument development. I believe that an instrument is not just culturally appropriate, it has to be grounded in the community as well. In this way, we see parents not as subjects, or even participants, but rather as partners who join with us in understanding children's socioemotional development.

Acknowledgments

The research described in this chapter was supported by the National Institute of Child Health and Human Development (HD 30590). I am indebted to my students and to the community service centers and research participants who over the years have helped me discern the meaning of cultural and community sensitivity.

References

Barrett, K. C., and Morgan, G. A. "Continuities and Discontinuities in Mastery Motivation During Infancy and Toddlerhood: A Conceptualization and Review." In R. H. MacTurk and G. A. Morgan (eds.), *Mastery Motivation: Origins, Conceptualizations, and Applications*. Norwood, N.J.: Ablex, 1995.

Busch-Rossnagel, N. A. "Commonalities Between Test Validity and External Validity in Basic Research on Hispanics." In K. F. Geisinger (ed.), *Psychological Testing of Hispanics.* Washington, D.C.: American Psychological Association, 1992.

Busch-Rossnagel, N. A. "Mastery Motivation in Toddlers." *Infants and Young Children,* 1997, *9,* 1–11.

Busch-Rossnagel, N. A., Knauf-Jensen, D. E., and DesRosiers, F. S. "Mothers and Others: The Role of the Socializing Environment in the Development of Mastery Motivation." In R. H. MacTurk and G. A. Morgan (eds.), *Mastery Motivation: Origins, Conceptualizations, and Applications.* Norwood, N.J.: Ablex, 1995.

Busch-Rossnagel, N. A., Vargas, M., and Knauf-Jensen, D. E. "Parenting Toddlers: Puerto Rican and Dominican Perspectives." In N. A. Busch-Rossnagel (chair), "Normative Research on Ethnic Minority Children." Symposium conducted at the meeting of the Society for Research in Child Development, Indianapolis, Ind., Mar. 1995.

Busch-Rossnagel, N. A., Vargas, M., Knauf, D. E., and Planos, R. "Mastery Motivation in Ethnic Minority Groups: The Sample Case of Hispanics." In D. Messer (ed.), *Mastery Motivation in Early Childhood: Development, Measurement, and Social Processes.* London: Routledge, 1993.

Combs, T. T., and Wachs, T. D. "The Construct Validity of Measures of Social Mastery Motivation." In D. Messer (ed.), *Mastery Motivation in Early Childhood: Development, Measurement, and Social Processes.* London: Routledge, 1993.

DesRosiers, F. S., and Busch-Rossnagel, N. A. "Self-Concept in Toddlers." *Infants and Young Children,* 1997, *10,* 15–26.

DesRosiers, F. S., and others. "Assessing the Multiple Dimensions of the Self-Concept of Young Children: A Focus on Latinos." *Merrill-Palmer Quarterly,* 1999, *45,* 543–566.

Dichter-Blancher, T. B., Busch-Rossnagel, N. A., and Knauf-Jensen, D. E. "Mastery Motivation: Appropriate Tasks for Toddlers." *Infant Behavior and Development,* 1997, *20,* 545–548.

Harter, S. "Effectance Motivation Reconsidered: Towards a Developmental Model." *Human Development,* 1978, *21,* 34–64.

Knauf-Jensen, D. E., Busch-Rossnagel, N. A., and Morgan, G. "Designing Comparable Instruments: Using Decentering to Create a Spanish Version of the Dimensions of Mastery Questionnaire." Poster presented at the biennial meeting of the Society for Research in Child Development, Washington, D.C., Apr. 1997.

MacTurk, R., and others. "Social Mastery Motivation in Down Syndrome and Nondelayed Infants." *Topics in Early Childhood Special Education,* 1985, *4,* 93–109.

Marín, G., and Marín, B. V. *Research with Hispanic Populations.* Thousand Oaks, Calif.: Sage, 1991.

Morgan, G. A., Busch-Rossnagel, N. A., Maslin-Cole, C., and Harmon, R. J. "Individualized Assessment of Mastery Motivation: Manual for 15- to 36-Month-Old Children." Unpublished document. Fordham University, Bronx, N.Y., 1992.

Rovinelli, R. J., and Hambleton, R. K. "Index of Item-Objective Congruence." In L. Crocker and J. Algina (eds.), *Introduction to Classical and Modern Test Theory.* New York: Holt, Rinehart & Winston, 1986.

Triandis, H. C. "Cross-Cultural Studies of Individualism and Collectivism." *Nebraska Symposium on Motivation,* 1989, *37,* 41–133.

Werner, O., and Campbell, D. T. "Translating, Working Through Interpreters and the Problem of Decentering." In R. Cohen (ed.), *A Handbook of Methods in Cultural Anthropology.* New York: American Museum of Natural History, 1970.

NANCY A. BUSCH-ROSSNAGEL is professor of psychology and dean of the Graduate School of Arts and Sciences at Fordham University, Bronx, New York.

6

A framework and some practical examples are presented for using rigorous implementation research to inform program outcomes and foster program development for developmental interventions.

Adherence Process Research on Developmental Interventions: Filling in the Middle

Aaron Hogue

Developmental interventions are programmatic efforts to influence change in specifically targeted developmental-behavioral processes and thereby alter the developmental trajectories of individuals or groups. Briefly stated, developmental interventions are *developmental* to the degree that they define individual behavior as a product of organism-context interactions over time, account for the organizational balance between risk and vulnerability factors on one hand and protective and resiliency factors on the other in developmental change, and assess the full profile of adaptive and maladaptive responses to developmental challenges at various life stages (Shirk, 1999). They are *interventions* to the degree that they promote change from an established or projected developmental pathway toward a more developmentally competent proximal or distal outcome (Ramey and Ramey, 1998).

Developmental interventions can be categorized according to three fundamental intervention goals. *Developmental health promotion* focuses on facilitating positive outcomes in normative developmental achievement across the life span (Zeldin, 2000). These interventions work by providing social environments that afford task performance or mastery and stimulate existing developmental competencies. *Preventive intervention* aims to prevent or delay the onset of psychological distress and mental health problems for all members of a given population (universal preventions), members who have empirically established risk characteristics (selective preventions), or members who exhibit subclinical symptoms of a psychological disorder (indicated preventions) (Mrazek and Haggerty, 1994).

Treatment intervention focuses on ameliorating symptoms and enhancing coping in individuals who experience significant psychological distress or exhibit behavioral symptoms that meet diagnostic criteria for mental health disorder (American Psychiatric Association, 1994).

The Role of Process Research: Developing Developmental Interventions

Assessing program effectiveness is an essential activity of developmental intervention research. However, studies have focused primarily on the front end of evaluation, participant recruitment and baseline assessment, and on the back end, outcome effects and follow-up assessment (Shonkoff, 2000). As a result, there is a significant dearth of empirical knowledge in the middle ground—program implementation and the immediate impact of programs on participants. The assessment of what occurs during the course of implementing a program is generally known as process evaluation. Process evaluations ask not whether but how programs produce effects (Judd and Kenny, 1981; Scheier, 1994). Note that process research can also be used to develop and pilot intervention programs or program components prior to formally implementing them. This kind of process research is often called *formative evaluation* (Scheier, 1994).

Psychotherapy is one intervention field with a long tradition of research interest in intervention process (Orlinsky, Grawe, and Parks, 1993). The main scientific lessons to be learned from psychotherapy's experiences with process research fall into two rough categories: method and utility. With regard to method, a schematic summary of developmental intervention process is presented in Figure 6.1. The intervention process is considered to be a collateral endeavor consisting of two interwoven elements: participant change and intervenor-participant exchange (see also Orlinsky, Grawe, and Parks, 1993).

Participant change refers to the developmental course of participant functioning in one or more behavioral areas during program delivery. Evaluation of participant change might occur at any point during the program, and immediate program impact can be measured within a single meeting or between two or more meetings. *Intervenor-participant exchange* refers to those aspects of intervention whereby the intervenor interacts directly with the participant in delivering the program. Note that this schematic of intervention process applies only to programs that entail personal delivery of intervention services—that is, to programs that use an intervenor or group of intervenors. It is therefore not applicable to programs that rely exclusively on videotaped protocols, direct mailings, bibliotherapy techniques, and the like. For heuristic purposes, this element of intervention process can be divided into three components: intervention type, intervenor skill, and relationship factors (see also Kazdin, 1993). *Intervention type* comprises both intervention parameters and intervention content and

Figure 6.1. Collateral Elements of Developmental Intervention Process

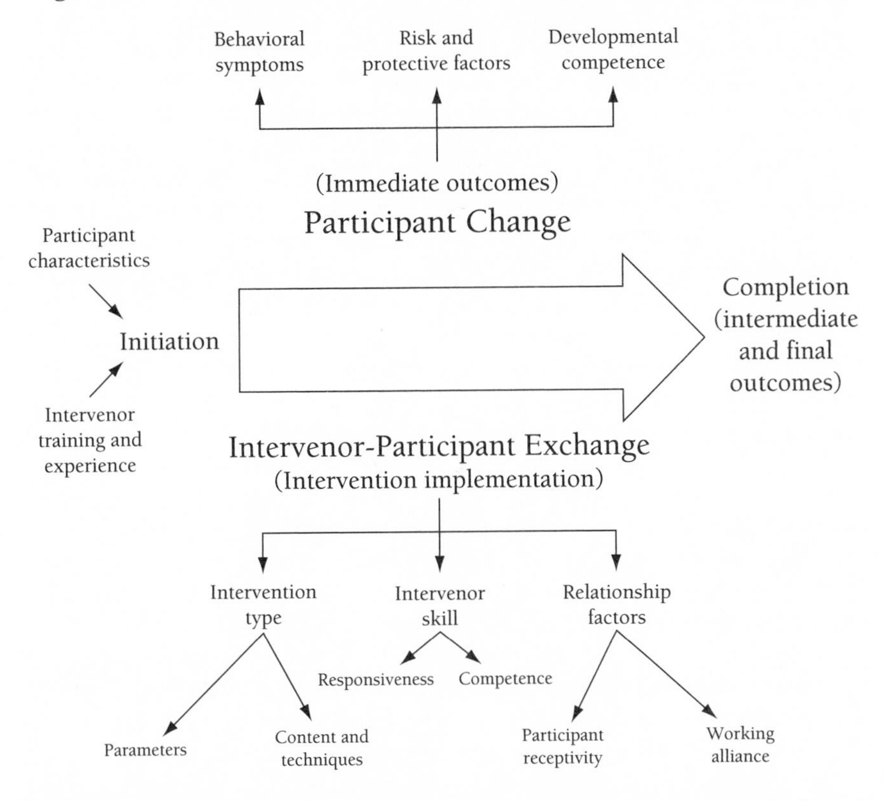

techniques. *Intervenor skill* includes responsiveness to participant behavior and competence in executing program goals. Finally, *relationship factors* include the broad spectrum of interpersonal interactions between intervenor and participant, especially participant receptivity and working alliance.

Process research is aimed at linking aspects of intervenor-participant exchange to targeted participant change both during the program and following its completion. Thus the ultimate utility of process research derives from its multifaceted role in assessing the immediate and cumulative impact of program implementation on developmental changes in participants. Process questions can yield answers that inform a host of issues related to program effectiveness, including intervenor training, mechanisms of program impact, intervenor and participant behaviors that shape outcome, and participant characteristics that predict differential responses. The common thread that binds these various functions of process research is intervention development (Gaston and Gagnon, 1996; Kazdin, 1993). By providing local (that is, program-specific) evidence of how an intervention produces its effects (or lack thereof), process research enables program developers to systematically review, critique, and revise key features of the intervention model. Given good process data, a program can only get better.

Adherence Process Research

Adherence process research is a comprehensive evaluation strategy used to examine one component of intervenor-participant exchange: intervention type (parameters, techniques, content). Specifically, adherence process research assesses program integrity—that is, the degree to which a given program is implemented in accordance with essential theoretical and procedural aspects of the intervention model (Hogue, Liddle, and Rowe, 1996). It has three identifying features. First, it uses quantitative measures to investigate the extent to which program parameters and techniques are implemented, including the intensity and frequency of specific intervenor behaviors. Second, it explores both program-specific intervention characteristics—elements of the theoretical model that are essential and perhaps unique—and generic characteristics endorsed by most programs (such as intervenor warmth and openness). Third, it considers how multiple intervenors and participants within a given program differentially influence the overall integrity of program delivery. That is, it considers how model, intervenor, and participant effects individually contribute to intervention process and outcome.

To produce detail-rich adherence process information, resource-intensive methods such as live or videotape observational ratings of program implementation are required. The quantitative nature of the data allows evaluators to link process assessment directly with program outcomes, thereby providing an avenue for investigating which parts of the program were most effective, and why (Gaston and Gagnon, 1996). Such information is critical not only for improving program efficacy but also for facilitating replicability and transportability to other settings and populations (Waltz, Addis, Koerner, and Jacobson, 1993). In short, adherence process evaluation surpasses the basic dichotomous judgment of most integrity checks—Was the program implemented as planned?—in favor of a multivariate assessment of intervention process—What occurred during program implementation?

Contributions of Adherence Process Research to Intervention

We have completed several adherence process studies in the service of theory building with a family-based developmental intervention for antisocial behavior: multidimensional family therapy, or MDFT (Liddle and others, 1992; Liddle and Hogue, 2001). MDFT is a multicomponent treatment for drug-using and conduct-disordered adolescents that works to change within-family interactions as well as interactions between the family and relevant social systems. MDFT is a highly flexible model that creates individualized treatment plans for each family, and it incorporates basic developmental research on adaptive versus maladaptive adolescent and family functioning into treatment planning designed to reduce or eliminate behavioral problems, repair family attachments, and foster a more prosocial developmental trajectory.

We first conducted adherence process research on the MDFT model for the purposes of model verification and calibration. Hogue and others (1998) compared intervention techniques exhibited by MDFT therapists to those exhibited by therapists practicing individual-based cognitive-behavioral therapy (CBT) during a randomized treatment study of adolescent substance abusers in outpatient treatment. The purposes of the adherence analyses were to verify treatment integrity and differentiation within the larger study and to discover what patterns of intervention techniques were emphasized by each set of therapists in the course of implementing flexible, multimodule treatments with a traditionally difficult-to-serve population.

Naive raters observed ninety videotaped therapy sessions across both models using an adherence process measure that tracks three kinds of therapist techniques: those uniquely endorsed by one of the models (level 1), those jointly endorsed by both models (level 2), and those commonly endorsed by most psychotherapies (level 3). We found that MDFT therapists reliably used the core systemic interventions uniquely prescribed by their governing model (level 1): shaping parenting skills, preparing for and coaching multiparticipant interactions in session, and targeting multiple family members for change. Also, MDFT therapists evidenced few individual differences in adherence. These results verified treatment fidelity within the MDFT condition.

In addition, we found that MDFT therapists employed certain jointly endorsed techniques to a significantly greater degree than their CBT counterparts (levels 2 and 3): establishing a supportive therapeutic environment, encouraging the expression of affect in session, engaging participants in setting a collaborative treatment agenda, and exploring themes related to normative adolescent development. Counter to expectations, family therapists were less likely to explore the details and ramifications of the adolescent's drug use behavior (level 2). These findings helped reshape our understanding of MDFT as an integrative model that works primarily in the attachment and affective domain. They also compelled us to consider the theoretical profits and hazards of comparatively underemphasizing drug use behaviors in favor of concentration on relationship issues. MDFT has since been revised to include directives for incorporating content related to current drug use behavior, as assessed by both self-disclosure and urine screens, into treatment sessions at regular intervals (Liddle, 2000).

We next used adherence process methods in the service of model specification. We adapted MDFT for use in a prevention context with young adolescents at risk for developing substance use and externalizing problems (Hogue, Liddle, and Becker, 2002; Liddle and Hogue, 2000). During the initial model-building phase of converting MDFT into a preventive intervention, multidimensional family prevention (MDFP), we delineated a foundation of

MDFT goals and techniques that would also anchor MDFP. In addition, we specified some intervention characteristics to be featured more strongly in MDFP than in MDFT, given the younger age range (eleven to fourteen years) and nonclinical status of the target population.

A total of 110 MDFP, MDFT, and CBT sessions were reviewed by a second group of observational raters (Hogue, Johnson-Leckrone, and Liddle, 1999). MDFP counselors were similar to MDFT therapists, and different from CBT therapists, in their use of signature techniques of multidimensional family intervention, such as encouraging the expression of affect and coaching family interactions in session. In addition, MDFP placed the greatest emphasis on enhancing family attachments and communication skills. Contrary to predictions, MDFP counselors were less likely than MDFT therapists to engage in many interventions that we had identified a priori to be primarily preventive in nature: discussing parental monitoring, helping families adopt a future orientation, and encouraging parents to become involved in the extrafamilial activities of their adolescents. In addition, a process examination of MDFP intervention parameters revealed that, in violation of their family-based practice guidelines, MDFP counselors spent as much time in sessions working alone with the adolescents as they did working with parents alone or with parent-adolescent dyads (Singer and Hogue, 2000). These findings underscore particular areas in which the MDFP model, and its training and supervision procedures, should be further articulated to serve high-risk prevention rather than clinical treatment populations.

A third adherence process study (Hogue, Samuolis, Dauber, and Liddle, 2000) was designed to explore process-outcome effects within the MDFT and CBT models. Outcome analyses from the randomized trial comparing MDFT and CBT found that the treatments produced equally successful outcomes at program completion in two critical domains of adolescent functioning: decrease in substance use, and decrease in externalizing behavior (Liddle, Turner, Tejeda, and Dakof, 2000). We then selected fifty sessions (twenty-six MDFT, twenty-four CBT) and identified two underlying process dimensions that captured both therapist technique and session content: *family focus* (for example, targeting family members for change, working on parental monitoring and family communication, focus on family issues) and *adolescent focus* (for example, building a working alliance with the adolescent, focus on peer issues, focus on drug use).

No process-outcome links were found for adolescent drug use behavior. However, across the two treatment conditions, level of family focus and level of adolescent focus each predicted outcome in externalizing behavior. Greater use of family-focused interventions predicted fewer conduct problems at posttreatment; in contrast, greater use of adolescent-focused interventions was associated with elevated conduct problems. These contrasting main effects for treatment process suggest that adolescents profited from more emphasis on family-based work and less emphasis on individual-based work. Follow-up analyses found no between-model differences for the negative effects of adolescent focus. However, the two models did receive substantively different benefits for use of family-focused techniques. Surprisingly, it was adolescents in CBT who were helped by greater inclusion of family techniques and issues in therapy; within MDFT, family focus did not predict externalizing outcomes. The overall implications for intervention development appear to be that, on balance, family-centered strategies are preferable to adolescent-centered strategies in reducing antisocial behavior in this population, and therapists applying individual-based models are well advised to integrate family themes into their treatment planning.

Conclusion

Developmental interventions can be greatly enhanced by rigorous evaluation of program implementation and of the diverse intervention processes by which programs achieve their effects. Adherence process research is a flexible methodological tool for assessing the nature and impact of intervention parameters, content areas, and techniques. As such, it plays a central role in the intervention development cycle that follows a course from theory development and program standardization through intervenor training and supervision to program delivery, and finally to evaluation of successes, failures, and surprises in implementing essential program elements.

Acknowledgments

Preparation of this chapter was supported in part by a grant from the National Institute on Drug Abuse (R03 DA12993). The author thanks Leyla Faw for her assistance in preparing the manuscript.

References

American Psychiatric Association. *Diagnostic and Statistical Manual of Mental Disorders.* (4th ed.) Washington, D.C.: American Psychiatric Association, 1994.

Gaston, L., and Gagnon, R. "The Role of Process Research in Manual Development." *Clinical Psychology: Science and Practice,* 1996, *3,* 13–24.

Hogue, A., Johnson-Leckrone, J., and Liddle, H. A. "Treatment Fidelity Process Research on a Family-Based, Ecological Preventive Intervention for Antisocial Behavior in High-Risk Adolescents." Paper presented at the annual meeting of the Society for Prevention Research, New Orleans, La., June 1999.

Hogue, A., Liddle, H. A., and Becker, D. "Multidimensional Family Prevention for At-Risk Adolescents." In T. Patterson (ed.), *Comprehensive Handbook of Psychotherapy,* Vol. 2: *Cognitive-Behavioral Approaches.* New York: Wiley, 2002.

Hogue, A., Liddle, H. A., and Rowe, C. "Treatment Adherence Process Research in Family Therapy: A Rationale and Some Practical Guidelines." *Psychotherapy: Theory, Research, Practice, and Training,* 1996, *33,* 332–345.

Hogue, A., Samuolis, J., Dauber, S., and Liddle, H. "Dimensions of Family Change in Multidimensional Family Therapy for Adolescent Substance Abuse." Paper presented at the annual conference of the American Psychological Association, Washington, D.C., Aug. 2000.

Hogue, A., and others. "Treatment Adherence and Differentiation in Individual Versus Family Therapy for Adolescent Substance Abuse." *Journal of Counseling Psychology*, 1998, *45*, 104–114.

Judd, C. M., and Kenny, D. A. "Process Analysis: Estimating Mediation in Treatment Evaluations." *Evaluation Review*, 1981, *5*, 602–619.

Kazdin, A. E. "Psychotherapy for Children and Adolescents." In A. Bergin and S. Garfield (eds.), *Handbook of Psychotherapy and Behavior Change.* (4th ed.) New York: Wiley, 1993.

Liddle, H. A. *Cannabis Youth Treatment (CYT) Manual*, Vol. 5: *Multidimensional Family Therapy Treatment (MDFT) for Adolescent Cannabis Users.* Rockville, Md.: Center for Substance Abuse Treatment, Substance Abuse and Mental Health Services Administration, 2000.

Liddle, H. A., and Hogue, A. "A Family-Based, Developmental-Ecological Preventive Intervention for High-Risk Adolescents." *Journal of Marital and Family Therapy*, 2000, *26*, 265–279.

Liddle, H. A., and Hogue, A. "Multidimensional Family Therapy for Adolescent Substance Abuse." In E. F. Wagner and H. B. Waldron (eds.), *Innovations in Adolescent Substance Abuse Interventions.* New York: Pergamon Press, 2001.

Liddle, H. A., Turner, R. M., Tejeda, M. J., and Dakof, G. A. "Treating Adolescent Substance Abuse: A Comparison of Cognitive-Behavioral Therapy and Multidimensional Family Therapy." Paper presented at the annual conference of the American Psychological Association, Washington, D.C., Aug. 2000.

Liddle, H. A., and others. "The Adolescent Module in Multidimensional Family Therapy." In G. Lawson and A. Lawson (eds.), *Adolescent Substance Abuse: Etiology, Treatment, and Prevention.* Gaithersburg, Md.: Aspen, 1992.

Mrazek, P. J., and Haggerty, R. J. (eds.). *Reducing Risks for Mental Disorders: Frontiers for Preventive Intervention Research.* Washington, D.C.: National Academies Press, 1994.

Orlinsky, D., Grawe, K., and Parks, B. "Process and Outcome in Psychotherapy: Noch Einmal." In A. Bergin and S. Garfield (eds.), *Handbook of Psychotherapy and Behavior Change.* (4th ed.) New York: Wiley, 1993.

Ramey, C. T., and Ramey, S. L. "Early Intervention and Early Experience." *American Psychologist*, 1998, *53*, 109–120.

Scheier, M. A. "Designing and Using Process Evaluation." In J. S. Wholey, H. P. Hatry, and K. E. Newcomer (eds.), *Handbook of Practical Program Evaluation.* San Francisco: Jossey-Bass, 1994.

Shirk, S. R. "Developmental Therapy." In W. K. Silverman and T. H. Ollendick (eds.), *Developmental Issues in the Clinical Treatment of Children.* Needham Heights, Mass.: Allyn & Bacon, 1999.

Shonkoff, J. P. "Science, Policy, and Practice: Three Cultures in Search of a Shared Mission." *Child Development*, 2000, *71*, 181–187.

Singer, A., and Hogue, A. "Therapist and Independent Observer Ratings of Therapist Adherence to a Family-Based Prevention Model." Paper presented at the annual meeting of the Society for Psychotherapy Research, Chicago, June 2000.

Waltz, J., Addis, M. E., Koerner, K., and Jacobson, N. S. "Testing the Integrity of a Psychotherapy Protocol: Assessment of Adherence and Competence." *Journal of Consulting and Clinical Psychology*, 1993, *61*, 620–630.

Zeldin, S. "Integrating Research and Practice to Understand and Strengthen Communities for Adolescent Development: An Introduction to the Special Issue and Current Issues." *Applied Developmental Science*, 2000, *4*, 2–10.

AARON HOGUE is assistant professor, Department of Psychology, Fordham University, Bronx, New York.

7

The idea that workplaces have levels of moral complexity is presented to assist in understanding how schools can create conditions that foster teachers' personal and professional growth.

The Necessity of Teacher Development

Ann Higgins-D'Alessandro

One purpose of applied developmental psychology is to conduct research about the interplay between societal institutions and the development, functioning, and well-being of individuals and groups. Society expects education as an institution—and teachers as individuals—to positively influence children, teaching them not only to read, write, and think in words and numbers but also to develop social and moral sensitivities, character, and citizenship. Yet the effect on teachers of the school context and job of teaching is often overlooked. Focus on the school environment and role of teaching can enhance the growth and development of both students and teachers, not only academically but also socially and morally. Some schools may enhance the way teachers teach and treat children but others may constrain and undermine them. When schools fail, society pins both its blame and its hope for change on teachers.

The Social and Moral Complexity of the Job of Teaching

The perspective advocated in this chapter emphasizes that teachers must continue to develop in order to ensure the intellectual, social, and personal development of children and youth, and that the organizational structures of schools can help to foster teachers' development. Accordingly, I present a conceptual framework for analyzing the effects of the job of teaching on the moral development of teachers. My original inspiration lies in the work of Melvin Kohn (1977) and Lawrence Kohlberg (1982, 1984).

Kohn (1977) developed a measure of intellectual challenge that he described as the substantive complexity of work. His longitudinal research showed that the level of substantive complexity of fathers' jobs related directly to the specific values parents chose to impart to their children. Fathers in jobs with little intellectual challenge promoted conformist values in their children, while those in challenging work encouraged individual autonomy. Parallel to this, I developed the concept of work moral complexity as a potential explanatory variable for understanding why some adults continue to develop in their moral reasoning while most do not. Therefore, I related the idea of work complexity to Kohlberg's stages of moral reasoning.

Theoretically, work conditions could create a ceiling for moral development. People reasoning at higher stages should be frustrated by the limitations of jobs defined by a lower level of moral complexity. Conversely, the moral reasoning of people using lower stages (stages 2 and 3) should be fostered by being in jobs with a higher level of moral complexity (levels 4 or 5).

Using this framework, I developed the Socio-Moral Complexity of Work (SMC-W) instrument, an individual qualitative structured interview and assessment measure. The SMC-W assesses eight social and moral domains of work in a wide range of jobs. The conditions of each domain are graded for increasing complexity across five levels. The five levels relate to Kohlberg's first five stages of moral development. SMC-W levels 3 and 5 appear in the appendix at the end of this chapter. (The SMC-W is one part of a larger project designed to analyze the conditions in schools that promote the development of students, teachers, and positive, sustained teacher-student relationships.)

In the late 1970s when Kohlberg and his colleagues reinterviewed his original sample, the fifty-eight boys were men in their forties. Six of the men served as the development sample for the SMC-W. This sample had two doctors, who scored stage 5, two lawyers, who scored stage 4, and two teachers who scored 3/4—the average adult moral reasoning stage in the United States—on the moral judgment interview (Heinz dilemma, and so on). Responses to additional questions about their lives indicated that job roles influenced and defined the subjects' sense of morality and values. The doctors were forced to think through moral issues during their daily decisions. The lawyers emphasized the negative impact of practicing corporate law on their personal values. The teachers felt the least allowance to effectively take moral responsibility when in situations with students or administrators.

This chapter uses the results of the SMC-W to illustrate the benefits of working in a school whose structures promote a high level of socio-moral complexity (level 5). Prior research indicates that in traditionally governed schools teachers report their jobs as level 3 socio-moral complexity (Higgins, 1995b).

Schools as Communities of Development

Just Community schools and programs are designed to promote the moral education of youth. The schools' organization also incorporates the development of teachers, by transforming the work of teaching into a complex social and moral experience.

The institutions of Just Community programs are structured to create moral experiences intermediate to the family and the more formal, adult relations we enjoy as colleagues, neighbors, and citizens. When the high school fosters the social and moral development of adolescents as well as their learning, it provides experiences of affiliative or caring relationships grounded in rule-governed and rule-following contexts (Power, Higgins, and Kohlberg, 1989). Its goal is the full development of students. To obtain this goal, John Dewey ([1938] 1963) as well as Kohlberg and Higgins (1987) emphasized that teachers should have sound knowledge of ethical and psychological principles—that is, they should be ethically mature themselves and able to advocate for the school as an ethical community.

Just Community programs exist in varying forms, including public school partial day programs, alternative public schools, and one private day school. They are defined by democratic governance, one-person-one-vote, student and teacher alike, and the space and number of members to enable large group face-to-face discussions. The high school model consists of a central governance structure: the weekly, two-hour community meeting in which all members participate, supported by smaller discussion groups (advisory groups), a fairness committee for resolving conflicts and rule violations, and an agenda committee. All groups and committees are made up of student and teacher members. Using the vehicles of weekly community meetings, advisory groups, and fairness and agenda committees, each Just Community program decides what issues are open for democratic decision making. Issues include rule making, rule violations, classroom management, peer relations, and community service. These issues make great discussions, since they so often focus on respect, fairness, justice, caring, and support. Curricular issues lie outside the democratic boundary, remaining the province of teachers' expertise and school and state regulations.

Teacher Development

The A School began in 1972 as an alternative high school in a suburb of New York City; by 1976 it was suffering from "free school malaise" (Power, Higgins, and Kohlberg, 1989, p. 194). Looking back, Tony Arenella, one of four founder-teachers and director for the past ten years, said, "Along comes Larry [Kohlberg] who has a way of drawing a circle around both of those claims, individualism and communitarianism. He suggests that we structure that conflict to promote growth. And all of a sudden we're feeling, instead of burnt out, like moral educators doing important work" (Codding and Arenella, 1981, p. 4).

Developing a Just Community school gave the teachers a sense that they were growing along with their students. When I interviewed him in 1987, Tony described how he had changed:

> It's going to be almost impossible for me to distinguish this school from important changes of personal growth in my life. It has got me a lot more in touch with myself as a human being. . . . It's the kind of school where you can't really hide behind a role. . . . In fact, conventional teaching was really getting to me. That's one of the reasons why we started this school.

In stark contrast, that same year a teacher from a large NYC high school seemed resigned to the constraints he felt in his job: "I think intellectually, or in any real way, I have not been that stimulated in the past twelve years even though I feel teaching is a big part of my life and I spend a lot of time doing it" (interview with author, 1987).

In the A School, Tony saw the necessary relationship between being a skilled teacher and being an ethically mature person, Dewey's ([1938] 1963) requirement: "I'm not discounting the [teaching] skills part. The fundamental thing has to do with one's own integrity, sense of values, one's own being, if you will. That's what essentially counts. With that you need to learn how to make that effective in terms of the nuts and bolts of teaching" (interview with author, 1987).

The interviews with A School teachers show that they all developed in their moral reasoning from stages 3 and 3/4 in 1987 to stage 5 in 1994 (Higgins, 1995a), a rare occurrence in any sample (Colby and Kohlberg, 1987). The founder-teachers described their teaching in level 5 terms at all three interview times, 1987, 1994, and 1999. In 1987 they described the level 5 complexity of their jobs created by the Just Community structures as challenging. By 1994, the continuity between their moral reasoning stage and their views of the moral complexity of teaching showed in the initiatives they suggested to foster the school community's development. I will use these teachers' more recent 1999 interviews to illustrate their continuing moral growth and indicate how the conditions of teaching that promote adult development are assessed by the SMC-W instrument.

The longitudinal results of the SMC-W interviews and assessments of the Just Community teachers indicate that a school can foster adult development if it provides the following four opportunities for teachers: (1) the opportunity to exercise their autonomy in teaching, curriculum, and schooling matters in the context of reasonable and contractual obligations; (2) the opportunity to actively participate and exercise power in school governance; (3) the opportunity to empathically take roles not only with their students but also with their colleagues, administrators, parents, and community members; and (4) the opportunity to take responsibility for creating fair processes and outcomes when conflicts arise. When the job of teaching comprises all these opportunities as well as experiences across the eight domains at level 5 (presented in the appendix), then the moral function of teaching becomes mentoring—that is, being responsible for the full development of each individual student, intellectually and personally.

The opportunity to exercise autonomy in teaching and curriculum in the context of contractual obligations (SMC-W domain: Evaluation). Sally Cohn, who has been the A School Spanish teacher for eight years, carefully exercises her autonomy in developing her curriculum and coursework, demonstrating level 5:

> I remember when I first came here I spoke nothing but Spanish and the students panicked. But it was really a dilemma for me because I didn't want the students to feel like they weren't learning, and yet I knew they were. And so I spoke to some of the other teachers in the school about it and I arrived at a compromise so the students would feel comfortable and feel that they, at the end of the day, walked away with something in their cup.

Many consider exercising such autonomy to be the mark of a good teacher; however, research shows that teachers are often not given that autonomy. In the urban high school, teachers said they prepared class notes according to district mandates but did not use them, keeping them in their desk drawers to demonstrate compliance. They felt the required district procedures constrained and hindered their teaching. Such conditions illustrate level 3 Evaluation, having to respond to external criteria. Their cynicism in interviews revealed the cost to these teachers.

The opportunity to actively participate and exercise power in school governance (SMC-W domains: Dialogue and Participation). Every A School teacher emphasized that participating in school governance was a responsibility as well as an opportunity. During a faculty meeting in spring 1999, the group discussed how their various roles as teacher, facilitator, advocate, and community planner intertwined to maintain the A School's democratic educational community. The issue was how to have a productive community meeting discussion if the students felt as if there were an alliance among the teachers. Tony Arenella put it this way:

> There are times when we all fall into the community point of view and students say, you know, the teachers are trying to get us to do something. We are. When that happens, it doesn't necessarily mean we have to retreat from that, but I think we have to be up front and say why we're all feeling the same—because we think it would be the best thing for the community—and we are speaking from that point of view and really not keeping it under cover, but being honest. It confuses them a little but otherwise they may simply vote against the argument because all the staff agrees.

Tom Conrad added,

> I think this advocate versus facilitator issue with an adjustment is important because you really have to switch in and out of those roles a lot.

Paying careful attention to their roles and to the probable structure of community meeting discussions are as important for the teachers as the issues.

Opportunities for empathic role-taking (SMC-W domain: Role-Taking). It's clear that empathic role-taking is demanded by the level 5 structures of a Just Community school and that they promote the growth of teachers as well as students. Howard Rodstein, a teacher at the A School for only one year said,

> I think I'm able to look at more than one side of an issue. . . . I think I really now try to analyze—what is the other guy thinking and how will they be affected by my thoughts? Also, I think it's brought home some of the emotional parts of the various arguments that we have here. Here, you look at how people feel and I think that's important for me. It really changed me. I really feel I've entered a great process of growth.

Tom, who has been the A School science teacher for nineteen years, discussed the intended effects on students when teachers are empathic. He said they encourage students to be both active and cooperative, to develop "the skills and talents they seem to inherently possess and some constructive pursuit," and "to participate in family life and kind of carry their weight there." He continued, "I think if all those things occur then they'll be able to avoid some of the destructive behavior in adolescence."

Opportunities to take responsibility for creating fair processes and outcomes when conflicts arise (SMC-W domains: Moral Conflict and Responsibility). In the 1999 interviews, the A School teachers generated the same kind of conflict and took responsibility in similar ways. The moral conflicts were whether to reveal to a parent what a student had said in confidence. The specific cases ranged from smoking to serious truancy to the fear of pregnancy. The teachers framed these as dilemmas of confidentiality and trust versus the student's welfare and the teachers' obligations to parents. They discussed these dilemmas and worked out a shared understanding of what to do, which included talking with the students about the decision to inform the parents, informing the parents, and being supportive of the students themselves but not necessarily of their positions or actions.

The Effects on Students and School Culture

In 1987, Tony Arenella reported an effect on the students that remains the teachers' cornerstone of success—students' reflecting on their own behavior and their relations with others.

> Kids think more about their actions in this program—it's built into the structure—than they're ever called upon to think about them in the more traditional sense, and to think about the actions of others. In other words, that

whole thing has become clearly conscious in this program. And to that extent, I think kids' consciousnesses are raised in that regard and they are much less likely to be passive people later.

Recently, as part of her research to understand adolescents' motivations for acting prosocially in school and with their peers, Lisa Markman, a Fordham University applied developmental psychology doctoral student, administered the School Culture Scale (Higgins-D'Alessandro and Sadh, 1997) to the A School faculty and students. Both teachers and students characterized student-student and student-teacher relationships more positively than did a group of comparison high school students. Moreover, they perceived norms against cheating and other problem behavior to be violated much less often, and they rated their educational experiences more highly as well.

Another important result of Markman's (2001) research demonstrates that positive experiences shared with teachers in school influence students' self-conceptualization as moral persons. In both the comparison group and the A School group, students who actively engaged in community service programs rated striving to meet their moral ideals as their primary motivation for many school-related behaviors (such as studying, doing homework, class participation). Thus, linking a prosocial or ethical activity with positive student-teacher relationships in the context of a learning environment encourages students to think of themselves in new, more ethically complex ways.

Conclusion

Schools have the resources to be communities of children and adolescents engaged with adults in shared tasks and projects that encourage both teacher and student development. This chapter illustrates the effectiveness of creating a school that is a community with conscious effort given to structuring education as an ethical and intellectual enterprise. In addition, the job of teaching is elevated to a high level of social and moral complexity, fostering teachers' own development. In such schools teachers are motivated to sustain positive relationships with children and youth and to be engaged with each other and the school, creating a positive spiraling relationship between one of society's most important institutions—education—and individual development.

Appendix: The Socio-Moral Complexity of Work

Level 3

Moral Conflicts: most often are considered to be peripheral to the job itself. They involve the claims/interests of the self with those of another individual or group on the job or with compliance to organization rules or authorities.

Role-Taking: is at times empathic and at other times instrumental or strategic. Empathic role-taking: may be expressed to clients or co-workers outside formal job parameters. Instrumental role-taking is done in relation to authorities and co-workers to make the job easier or be successful.

Dialogue: about welfare and fairness issues is not required and often is not allowed for these jobs by organization rules or policies. If dialogue is used to solve work-related issues among co-workers, it may not be legitimated by authorities.
Responsibility: for fair or responsible outcomes is not required and often is not allowed. Instead responsibility is defined as doing one's job, so that the next person can do her/his job, thus maintaining a chain of responsibility and authority.
Participation: in policy-making is allowed only to the extent given by authorities or normal procedures, e.g., grievance procedures negotiated between a union and a company.
Evaluation: standards are external criteria for job performance which may or may not be public and objective.
Constituencies: are people directly affected by one doing his/her job, primarily one's clients and immediate supervisors.
Moral Function: is expressed as doing one's job as defined by higher authorities, policies and rules, and laws.

Level 5
Moral Conflicts: are considered central to the job, are of high complexity, give priority to moral claims, and balance competing claims of fairness and welfare of individuals, groups, and/or the organization as a whole.
Role-Taking: is only empathic; rights and welfare of others are the focus and all people associated with a situation are included in one's role-taking.
Dialogue: about welfare and fairness issues is defined as part of one's job. Dialogue is open and attempts to include all people affected.
Responsibility: for fair and responsible outcomes is defined as part of one's job. Failure to reach just or responsible solutions is understood as failing to do one's job.
Participation: in policy making is not set only by rules, but is an obligation of the person in the job, who must join with others to create policies and to evaluate them in light of the moral, social and other goals of the organization.
Evaluation: standards are self-chosen and derive from one's view of the ideal conception of the job. These include but go beyond the objective criteria for job performance that the person has accepted as legitimate.
Constituencies: all people affected by situations and decisions on the job are considered, whether they are within or outside the organization, even including future generations. People's claims and interests are also considered in light of group memberships and individual histories.
Moral Function: at this level, the core moral goal or definition of the job is articulated and the rest of the criteria are coordinated in terms of it. Example: The moral term for teacher is mentor, which includes both the normative component of being responsible for and the value component of respect for the inherent dignity and potential capacity of every student.

References

Codding, J., and Arenella, A. "Creating a 'Just Community': The Transformation of an Alternative School." *Moral Education Forum,* 1981, 6(4), 2–7.
Colby, A., and Kohlberg, L. *The Measurement of Moral Development.* Vol. 1. New York: Cambridge University Press, 1987.
Dewey, J. *Experience and Education.* New York: Collier, 1963. (Originally published 1938.)

Higgins, A. "Educating for Justice and Community." In W. Kurtines and J. Gewirtz (eds.), *Moral Development: An Introduction*. Boston: Allyn & Bacon, 1995a.

Higgins, A. "Teaching as Moral Activity: Listening to Teachers in Russia and the United States." *Journal of Moral Education*, 1995b, *24*, 143–158.

Higgins-D'Alessandro, A., and Sadh, D. "The Dimensions and Measurement of School Culture: Understanding School Culture as the Basis of School Reform." *International Journal of Educational Research*, 1997, *27*, 553–569.

Kohlberg, L. *Essays on Moral Development*, Vol. 1: *The Philosophy of Moral Development*. San Francisco: Harper San Francisco, 1982.

Kohlberg, L. *Essays on Moral Development*, Vol. 2: *The Psychology of Moral Development*. San Francisco: Harper San Francisco, 1984.

Kohlberg, L., and Higgins, A. "School Democracy and Social Interaction." In J. Gewirtz and W. Kurtines (eds.), *Social Development and Social Interaction*. Somerset, N.J.: Wiley-Interscience, 1987.

Kohn, M. *Class and Conformity: A Study in Values*. (2nd ed.) Chicago: University of Chicago Press, 1977.

Markman, L. "The Impact of School Culture on Adolescents' Prosocial Motivation and Behavior." Unpublished doctoral dissertation, Fordham University, Bronx, N.Y., 2001.

Power, F. C., Higgins, A., and Kohlberg, L. *Lawrence Kohlberg's Approach to Moral Education*. New York: Columbia University Press, 1989.

ANN HIGGINS-D'ALESSANDRO is associate professor and director of the Graduate Program of Developmental Psychology at Fordham University, Bronx, New York.

8

Seven points are presented, using research to examine research-policy connections, to illustrate the role of psychology in policymaking.

The Role of Psychological Research in Setting a Policy Agenda for Children and Families

Lonnie R. Sherrod

This volume describes the stories of the lives of four psychologists who have been influential in using research to inform social policy. In this chapter, I argue first that not only is it appropriate for psychologists to concern themselves with policy, it is essential that attention to policy become an important aspect of psychological research. Second, I describe some principles that are important to using research to guide our attention to policy. It is critical that psychological and child developmental research play a role in the policymaking process if we are to develop the best programs and policies to serve children and families.

The Case for Psychological and Child Developmental Attention to Policy

An applied perspective was important in the early history of both psychology and child development, but this concern for application was somewhat lost during the empiricist tradition of the learning theorists who dominated the field in this country during mid-century. Whereas even Skinner had the idea of using his behaviorist ideas to build a utopian society—an applied interest—the concern with methods, particularly experimentation, led to a focus on laboratory research that became increasingly divorced from the real world. It is only in the past few decades that psychologists have recognized the importance of context, of qualitative methods, and of secondary analyses of large datasets for a full understanding of the topics they investigate

(Brooks-Gunn, Berlin, Leventhal, and Fuligni, 2000). This broadening of perspective has also increased the relevance of psychological, including child developmental, research to the social problems faced by society, particularly children and families.

First, it is essential that psychologists be present at the policymaking table. Much policy is oriented to changing human behavior, whether reducing high-risk sexual behavior, preventing teen pregnancy, improving school performance, or reducing substance abuse. It is unfortunate that our policymaking has been disproportionately driven by economists and scholars other than psychologists who study behavior. If we are to design policies that improve the lives of people, it is critical that psychologists who understand behavior and developmentalists who study children and families be involved.

Second, policies and programs are part of the surround in which people live and hence are critical areas for psychological research. I have elsewhere argued for the importance of studying programs as contexts for youth development (Sherrod, 1997). Bronfenbrenner's ecological systems theory (Bronfenbrenner and Morris, 1998) describes the importance of macrosystem variables of culture and social norms and exosystem variables involving social institutions—as well as the mesosystem, family and schools, and the microsystem of the individual child. It is time that policies and programs are included as part of the mesosystem and macrosystem as social institutions influencing socialization and development. From this perspective, policy work is not substantially different from clinical attention; it is just focused at a macro level rather than trying to solve individual problems. Psychology programs such as the Applied Developmental Program at Fordham University, which sponsored the Influential Lives Conference, do this as part of their mission, but the importance is too great to be limited to a few select programs. Policymaking should be as core to our field as research methods and statistics. If we are to develop effective policies, it is essential that psychologists be involved, and policies and programs provide important areas for psychological research.

Building Policy and Research Connections

To some extent, it is easier to involve psychology in the policymaking process than it is to include research. Although psychology is a science, it also has a practitioner element that increases its relevance to policy. Clinical, forensic, and industrial-organizational are applied branches of psychology, although they tend to function at an individual rather than a systems level. And as I have argued, policymaking should be added to this array (Sherrod, 1997). Although it should be obvious that information from research should be useful to policymaking, that usefulness is, in fact, too frequently not recognized. First, other factors such as ideology or cost of the policy may reduce the impact that research can have on the policymaking process.

Second, it is frequently difficult for research to provide the clear, direct, singular answer-type guidance that is needed for policymaking. Third, social problems change faster than our ability to generate information to address them (Prewitt, 1995). Hence there is relatively constant pressure against using research to guide policy. As the contributors to this volume make evident, the need to base policymaking in research must be always on the agenda of the applied researcher.

I will use concrete examples to highlight the importance of building and maintaining connections between research and policy, especially today when such serious social problems confront children, youth, and families. I hope the seven points I make will offer some guidance to strengthening this connection as well.

For several years, I have worked with the Committee on Child Development, Public Policy and Public Information of the Society for Research on Child Development (SRCD)—the policy arm of SRCD. During this same period, I was vice president of the William T. Grant Foundation, a private independent foundation dedicated to funding research on children and youth and by its mission concerned about increasing the usefulness of research for policymaking. The ideas presented here are based on these experiences. In fact, most researchers—like me, and like the other authors of this volume—find their interest in policy through some indirect route, since psychology programs did not and even now do not devote much attention to policy. However, my impression, having worked with students on the SRCD Committee, is that the younger generation is very interested in research-policy connections; we must exploit this interest by developing institutional mechanisms for young scholars to follow the type of career paths represented by the researchers showcased in this volume.

I wish to make seven points about developing and maintaining a close interaction between research and policy. Each point is illustrated with data and policy regarding a particular social problem.

1. *It is necessary to use both demographic information summarizing a problem and research findings that address the underlying causes and consequences of the problem.* Research on and policy attention to violence demonstrate this point. Demographic information documents the importance of violence as a social problem. Research points to some reasons why current practices may not have been effective and to new directions for policy attention.

Although there may currently be an epidemic of violent crimes, incarceration and punishment policies as used in this society do not seem to deter violence. The violent crime rate has increased substantially across the past decade, and as a result, so has our incarceration rate. In fact, the violent crime rate and the incarceration rate in this country increase at the same level; they show almost identical trajectories (Hamburg, 1995), thus we need a different strategy. Research information points to some new approaches to the problem.

Youth crime is in large part responsible for the growth in crime; the violent crime rate for youth increased across the past decades while remaining level for other age groups (Snyder and Sickman, 1995). Violence rises in adolescence, and typically drops off in young adulthood.

Adolescence is a time of experimentation with many adult behaviors. Part of our responsibility as parents and as a society is to help protect adolescents from long-term damage during this period of experimentation and vulnerability (Hamburg, 1995). Several longitudinal studies now show that if adolescents are protected from damage they eventually get back on their feet (Werner and Smith, 1992).

There has been little change in the types of delinquent acts adolescents report committing (Bachman, Johnston, and O'Malley, 1993). However, as might be expected, most crime occurs in the hours after school, from three to eleven in the evening. Activities aimed at occupying these hours for youth through clubs, community organizations, schoolwork, and the like would go a lot further toward preventing crime than waiting for adolescents to commit crimes and then incarcerating them (Task Force on Youth Development and Community Programs, 1992).

In addition, there is a sharp drop-off in violent crimes from adolescence into young adulthood for most youth (Elliot, 1994). The one exception is young black males. Young black males have the highest rate of violence and incarceration; they are the only group who show continuity of adolescent levels of violence into young adulthood. However, for young black males with a steady relationship or stable job, the curve looks like that for whites and girls (Elliot, 1994); that is, part of the reason for staying with a career of crime appears to be the lack of a successful transition to adulthood. Therefore, one solution to the violent crime rate is to help youth, particularly African American males, make a successful transition to adulthood by providing education, training, and jobs.

2. *Both basic and applied research are needed to inform policy*. Furthermore, the applied category can be differentiated as policy-relevant research and policy analysis. All three types of research are needed in our approach to poverty. Poverty, and its related issues such as lack of access to good health care and child care, is probably the most important issue facing children and families. Poverty disproportionately affects minorities, and fuels those factors such as family violence and substance abuse that relate to youth violence.

Children are disproportionately poor and the U.S. stands out in this regard. The impact of poverty on children is one important area of policy-relevant research. One route to the impact of poverty on children is poverty's impact on parents' behavior. Poverty stresses parents and undermines their successful parenting; there are excellent examples of policy-relevant research that demonstrate the mechanisms through which poverty affects children's development (McLoyd, 1998). Basic research demonstrates what makes for successful, effective parenting; thus basic research provides a yardstick against which to apply the changes we see resulting from poverty (Baumrind, 1993). Poverty also affects neighborhoods, which in turn influence child development (Duncan, Brooks-Gunn, Young, and Smith, 1998). Policy solutions can be directed to breaking the multiple impacts of poverty at each point of origin, for example, through parent support to help poor parents be successful parents, through neighborhood improvement projects such as zones of empowerment or the new MTO project, aimed at moving families to better neighborhoods, or by simply giving them back resources as in earned income tax credits. The point is that both basic and policy-relevant research are necessary to understand the impact of poverty on children and to identify target points and strategies of intervention.

The role of income in poverty's effects is currently being tested in at least one intervention, and this example brings up the last point in this section, the importance of policy analysis, especially by psychologists who understand child development. The New Hope Project, a demonstration project based in Milwaukee, Wisconsin, aims to give poor people the resources they need to improve their lives rather than attempting to give them the skills and training they need to improve their situation themselves. The New Hope model seems to have more effects and more positive impact on children than the usual intervention model oriented toward "improving" the parents so they may get jobs. New Hope generated higher achievement and lower behavior problems in boys (Huston and others, 2001). This is the first systematic attempt I have seen to directly compare policy solutions to a social problem such as poverty. A strength of this project is its use of psychologists who know how to assess child development.

We need a whole new generation of policy analysis done by psychologists who are familiar with and sensitive to child development. Research using survey data with large national databases has typically shown effects of poverty only during early childhood and only for cognitive variables, not for social-emotional variables such as problem behavior (Duncan, Brooks-Gunn, Young, and Smith, 1998). I find this result to be surprising, and it may result from the limited information on children available in survey items. We may need some measures other than surveys to see the full magnitude of effects on child development.

There is a noteworthy example carried out by both Manpower Demonstration Research Corporation (MDRC) in its evaluation of New Chance, a teen mother demonstration project, and Child Trends in its evaluation of JOBS, the parental training program associated with the Family Support Act of 1988. Measures used by Snow and Egeland in their child developmental research were added to the evaluations of these programs. These measures showed effects on the parents of increasing the mean length of utterances during reading and reducing harsh treatment of children. Both of these parental behaviors have been shown through basic research to be important to child outcomes (Zaslow and Eldred, 1998). Hence, positive child outcomes should emerge through the impact of the changed parental behavior. This type of applied work using methods and measures known to be revealing about child development will help us understand better the impact of poverty and of antipoverty programs on children.

3. *Developmental appropriateness and developmental continuity are crucial considerations.* The latter point means that there are no magic bullets and the former emphasizes the role of developmental research. Early development, brain growth, and language learning are now popular topics in the public press, due in part to an important report by the Carnegie Corporation (1994); these topics illustrate the importance of developmental appropriateness and developmental continuity to policy. I agree that the first few years are important, but so is subsequent development, and all is not lost if we miss the opportunity in these first few years (Sherrod, 1997). The area of early intervention highlights the importance of developmental appropriateness and developmental continuity to policy.

Life span researchers emphasize the importance of developmental embeddedness (Lerner, 1998). This idea means that each period of the life span is embedded in the periods that came before and will come after. In attending to one period, such as infancy or adolescence, we have also to attend to what happened in the previous period (prenatally or during school age) and address the period with an eye to consequences during the next period of life (toddlerhood or young adulthood). This orientation applies to policy just as it does to research. Applied developmental psychology proposes the idea of temporality of change, meaning that a view of the life span should be taken when evaluating intervention programs. It would be inappropriate to expect that an intervention designed to increase focused attention in infants will have a continuing impact on development out of context and into more complex levels of cognitive functioning at later ages. No intervention can necessarily lead to lasting change beyond its focus (Fisher and Lerner, 1994).

Early interventions such as Head Start have typically been evaluated for both short-term and long-term changes. Neither evaluation strategy has been attentive to the ideas of developmental embeddedness and temporality of change. Evaluations have shown important impacts immediately after the Head Start experience in terms of readiness for school; longer-term evaluations have shown such effects generally fade out over time (Zigler and Hall, 2000). The presence of short-term effects demonstrates the success of this important program. This success is doubly demonstrated by the fact that most Head Start evaluations have focused on cognitive variables that were not the focus of the originators of the program.

Expecting long-term effects from a short-term program and focusing on a narrow range of outcomes are both evaluation strategies that are wrongheaded and reflect an ignorance of child development. To expect long-term effects, we must ask by what developmental mechanisms we would expect a preschool program that lasts for a year or two to generate lifelong effects. This frequently reflects an inoculation model of policy, which—while perhaps attractive to policymakers because it points to easy solutions—does not agree with our knowledge of developmental process. Preschool programs can set children on a good developmental path by increasing their readiness for school, but that effect can last only if they go to reasonable schools and receive some form of family support. A snowball effect has been proposed to explain how a short-term program such as Head Start might generate long-term effects (and a few have been reported), but a preschool program cannot alleviate the consequences of a family life in poverty or poor schools following the preschool. Given the life circumstances faced by most children who enter Head Start, it is not reasonable to expect a snowball effect from the program.

In evaluations of programs such as Head Start, we need some theory of change for how the intervention is to generate effects, whether short term or long term. An evaluation then becomes a test of this theory in regard to development (Connell, Aber, and Walker, 1995). Furthermore, even if preschool is succeeded by at least average experiences that allow the child to continue to progress, some type of booster interventions may be needed to maintain the effects of the early experience (Zigler and Hall, 2000). Finally, we also need to adopt more of a cumulative view of the impact of programs across a child's life span. From this view, one might ask how participation in Head Start, followed by Head Start transition programs, followed by after-school programs during the school years, and by youth development programs such as Boys and Girls Clubs, could generate cumulative outcomes in regard to young adulthood status (Sherrod, 1997).

4. *It is essential that we adopt a multifaceted perspective in the design of policies and in their evaluation and assessment.* This has been demonstrated in the youth field, which has moved from treatment to prevention to promotion of positive development (Scales, Benson, Leffert, and Blyth, 2000). This is an orientation to policy that recognizes that all youth have needs, and youth vary not only in individual traits that make for success but in the extent to which their needs are met by the resources naturally occurring to them—in the form of family, schools, and communities. The goal of policy is to fill in the gaps in these resources. From this view of youth development, policies and programs also become part of the surroundings in which youth grow up.

This orientation raises a number of interesting ideas in regard to policy and its evaluation. First, it orients our attention to the environment as well as to the individual. Hence, policy should be directed to remedying deficiencies in the environment, not just to fixing individuals. We need also to do more needs assessment in order to determine the appropriate balance between fixing the environment and fixing the individual.

Huston and others (2001), in an ingenious analysis of antipoverty programs, noted that policy attention is usually oriented to fixing individuals. In contrast, the New Hope program, mentioned previously, attended to the environment of poor individuals and attempted to give them needed resources instead of education, skills, and so on, thus changing their environment.

Hence, while experimental evaluations that allow attribution of causality to a program are desirable, we also want to research programs and policies as contexts for development and to examine how program effects may cumulate over the life span.

5. *Dissemination is a key ingredient of the research-policy interaction, but the target audience of each dissemination must be clear and should vary with both the problem being addressed and the policy being proposed.* Research on dissemination is needed. Several years ago, Ellen Greenberger single-handedly prevented passage of an adolescent employment bill through her testimony before Congress on her research on the consequences of teen employment (McCartney and Phillips, 1993). In this case, the legislators in charge of enacting a law at that moment were the target. The Center on Children in Poverty at Columbia University, directed by Lawrence Aber, has identified the public as its target, and Gilliam and Iyengar (2000) have targeted the media's framing of news stories. Targeting several different audiences is critical for achieving a common goal—here, to improve policies for children and youth.

We do not know enough about the effectiveness of different dissemination strategies. Psychologists should build in research on dissemination efforts in the same way that we should build evaluations and research into program efforts. In addition, Thompson and Nelson (2001) have shown that dissemination of research results can be misleading. It is encouraging that many dissemination efforts by researchers are currently under way across the country (Sherrod, 1999).

6. *Cost-benefit analyses and recommended means of achieving costs must be part of the efforts to help children and families; otherwise, failure is assured.* It is imprudent to propose any policy effort without some idea of how to fund it. However, attention to cost-benefit analyses is not always as rational as one might like. We resist looking to other countries for advice and we seem incapable of adopting a long-term perspective. This country's orientation to parental leave is one example. We have bare-bones policy in the form of the 1993 Family and Medical Leave Act (FMLA) because of concern for how employers can bear the costs (Zigler and Hall, 2000). Cost to employers, especially small businesses, is a difficult issue, but solutions in other countries should offer some guidance (Kammerman, 2000). We also seem resistant to taking a long-term perspective on costs. There have been numerous reports documenting how early preventive efforts in regard to delinquency and crime would be much less costly than incarceration; yet we seem intent on maintaining a get-tough stance on crime. Researchers have an obligation to present the evidence for the effectiveness of a longer-term perspective.

7. *Research will be only one of the many factors driving policy.* Research practitioners have to recognize this fact and do the best they can.

Research and policy attention to child care illustrate clearly how research alone confronts an uphill battle in influencing policy. For example, there have been dramatic changes in the U.S. family from a two-parent farm family to the current dual career or single-parent family. The family has also shrunk in size as sibships have decreased and the extended family has disappeared (Hernandez, 1993). Given this change in the family, the issue of who is raising our children clearly arises (Hamburg, 1992).

Outstanding child development and child care research has demonstrated that we can define good-quality care that results in positive child development outcomes (Lamb, 1998). The issue now is how to implement the needed type of care in a cost-effective way. This good research, regrettably, has not been sufficient to generate the type of federal policy we need. At the local level it can be used to advocate for better policies. However, as valuable and necessary as research is, we cannot expect it to rule the day when it comes to policymaking.

Conclusion

I have argued for the importance of psychologists to attend to policy because we have important contributions to make. I have also outlined a few core issues in regard to policymaking and the role of research. There is a clear need in our field for more direct attention to the importance of the research-policy link, and I hope that this chapter will contribute to that goal.

References

Bachman, J., Johnston, J., and O'Malley, P. *Monitoring the Future.* Ann Arbor: University of Michigan Press, 1993.

Baumrind, D. "The Average Expectable Environment Is Not Good Enough: A Response to Scarr." *Child Development,* 1993, *64,* 1299–1317.

Bronfenbrenner, U., and Morris, P. A. "The Ecology of Developmental Processes." In W. Damon (gen. ed.) and R. M. Lerner (vol. ed.), *Handbook of Child Psychology,* Vol. 1: *Theoretical Models of Human Development.* (5th ed.) New York: Wiley, 1998.

Brooks-Gunn, J., Berlin, L., Leventhal, T., and Fuligni, A. "Depending on the Kindness of Strangers: Current National Data Initiatives and Developmental Research." *Child Development,* 2000, *71,* 257–268.

Carnegie Corporation of New York. *Starting Points: Meeting the Needs of Our Youngest Children.* Waldorf, Md.: Carnegie Corporation of New York, 1994.

Connell, J., Aber, J. L., and Walker, G. "How Do Urban Communities Affect Youth? Using Social Science Research to Inform the Design and Evaluation of Comprehensive Community Initiatives." In J. Connell, A. Kubisch, L. Schorr, and C. Weiss (eds.), *New Approaches to Evaluating Community Initiatives.* New York: Aspen Institute, 1995.

Duncan, G., Brooks-Gunn, J. B., Young, J., and Smith, J. "How Much Does Childhood Poverty Affect the Life Chances of Children?" *American Sociological Review,* 1998, *63,* 406–422.

Elliot, D. "Serious Violent Offenders: Onset, Developmental Course, and Termination." *Criminology,* 1994, *32,* 1–21.

Fisher, C., and Lerner, R. *Applied Developmental Psychology.* New York: McGraw-Hill, 1994.

Gilliam, F., and Iyengar, S. "Prime Suspects: The Impact of Local Television News on Attitudes About Crime and Race." *American Journal of Political Science,* 2000, *44,* 560–573.

Hamburg, B. "President's Report. The Epidemic of Youth Violence: Effective Solutions Require New Perspectives." *William T. Grant Foundation Annual Report.* New York: Fell, 1995.

Hamburg, D. *Today's Children: Creating a Future for a Generation in Crisis.* New York: Times Books, 1992.

Hernandez, D. *America's Children: Resources from Family, Government, and the Economy.* New York: Russell Sage Foundation, 1993.

Huston, A. C., and others. "Work-Based Anti-Poverty Programs for Parents Can Enhance the School Performance and Social Behavior of Children." *Child Development,* 2001, *72,* 318–336.

Kammerman, S. "Parental Leave Policies: An Essential Ingredient in Early Childhood Education and Care Policies." *Social Policy Reports,* 2000, *14*(2). Entire issue.

Lamb, M. "Nonparental Child Care: Context, Quality, Correlates and Consequences." In W. Damon (gen. ed.) and I. Sigel and K. Renninger (vol. eds.), *Handbook of Child Psychology,* Vol. 4: *Child Psychology in Practice.* (5th ed.) New York: Wiley, 1998.

Lerner, R. M. "Theories of Human Development: Contemporary Perspectives." In W. Damon (gen. ed.) and R. M. Lerner (vol. ed.), *Handbook of Child Psychology,* Vol. 1: *Theoretical Models of Human Development.* (5th ed.) New York: Wiley, 1998.

McCartney, K., and Phillips, D. *An Insider's Guide to Providing Expert Testimony Before Congress.* Ann Arbor, Mich.: Society for Research in Child Development, 1993.

McLoyd, V. "Children in Poverty: Development, Public Policy, and Practice." In W. Damon (gen. ed.) and I. Sigel and K. Renninger (vol. eds.), *Handbook of Child Psychology,* Vol. 4: *Child Psychology in Practice.* New York: Wiley, 1998.

Prewitt, K. *Social Sciences and Private Philanthropy: The Quest for Social Relevance.* Essays on Philanthropy, no. 15. Series on Foundations and Their Role in American Life. Indianapolis: Indiana University Center on Philanthropy, 1995.

Scales, P., Benson, P., Leffert, N., and Blyth, D. "Contribution of Developmental Assets to the Prediction of Thriving Among Adolescents." *Applied Developmental Science,* 2000, *4,* 27–46.

Sherrod, L. "Promoting Youth Development Through Research-Based Policies." *Applied Developmental Science,* 1997, *1,* 17–27.

Sherrod, L. "Giving Child Development Knowledge Away: Using University-Community Partnerships to Disseminate Research on Children, Youth and Families." *Applied Developmental Science,* 1999, *3,* 228–234.

Snyder, H. N., and Sickman, M. *Juvenile Offenders and Victims: A National Report.* Washington, D.C.: National Center for Juvenile Justice, 1995.

Task Force on Youth Development and Community Programs, Carnegie Council on Adolescent Development. *A Matter of Time: Risk and Opportunity in the Nonschool Hours.* New York: Carnegie Council on Adolescent Development, 1992.

Thompson, R., and Nelson, C. "Developmental Science and the Media: Early Brain Development." *American Psychologist,* 2001, *56,* 5–15.

Werner, E., and Smith, R. *Overcoming the Odds: High Risk Children from Birth to Adulthood.* Ithaca, N.Y.: Cornell University Press, 1992.

Zaslow, M., and Eldred, C. (eds.). *Parenting Behavior in a Sample of Young Mothers in Poverty: Results of the New Chance Observational Study.* New York: Manpower Demonstration Research Corporation, 1998.

Zigler, E., and Hall, N. *Child Development and Social Policy.* New York: McGraw-Hill, 2000.

LONNIE R. SHERROD is professor of psychology at Fordham University, Bronx, New York.

Index

Back Issue/Subscription Order Form

Copy or detach and send to:
Jossey-Bass, A Wiley Company, 989 Market Street, San Francisco CA 94103-1741

Call or fax toll-free: Phone 888-378-2537 6:30AM – 3PM PST; Fax 888-481-2665

Back Issues: Please send me the following issues at $28 each
(Important: please include series initials and issue number, such as CD99.)

____

____

____

$ __________ Total for single issues

$ __________ SHIPPING CHARGES:

SURFACE	Domestic	Canadian
First Item	$5.00	$6.00
Each Add'l Item	$3.00	$1.50

For next-day and second-day delivery rates, call the number listed above.

Subscriptions: Please __start __renew my subscription to *New Directions for Child and Adolescent Development* for the year 2_____at the following rate:

U.S.	__Individual $79	__Institutional $180
Canada	__Individual $79	__Institutional $220
All Others	__Individual $103	__Institutional $254
Online Subscription		__Institutional $180

For more information about online subscriptions visit www.interscience.wiley.com

$ __________ Total single issues and subscriptions (Add appropriate sales tax for your state for single issue orders. No sales tax for U.S. subscriptions. Canadian residents, add GST for subscriptions and single issues.)

__Payment enclosed (U.S. check or money order only)
__VISA __MC __AmEx __Discover Card #__________________ Exp. Date ______

Signature ______________________________ Day Phone ____________
__ Bill Me (U.S. institutional orders only. Purchase order required.)

Purchase order # ____
Federal Tax ID13559302 **GST 89102 8052**

Name ____

Address ____

____

Phone ______________________ E-mail ___________________________

For more information about Jossey-Bass, visit our Web site at www.josseybass.com

PROMOTION CODE ND03

Other Titles Available in the
New Directions for Child and Adolescent Development Series
William Damon, Editor-in-Chief

CD97 Talking Sexuality: Parent-Adolescent Communication, *S. Shirley Feldman, Doreen A. Rosenthal*

CD96 Learning in Culture and Context: Approaching the Complexities of Achievement Motivation in Student Learning, *Janine Bempechat, Julian G. Elliott*

CD95 Social Exchange in Development, *Brett Laursen, William G. Graziano*

CD94 Family Obligation and Assistance During Adolescence: Contextual Variations and Developmental Implications, *Andrew J. Fuligni*

CD93 Supportive Frameworks for Youth Engagement, *Mimi Michaelson, Anne Gregor, Jeanne Nakamura*

CD92 The Role of Family Literacy Environments in Promoting Young Children's Emerging Literacy Skills, *Pia Rebello Britto, Jeanne Brooks-Gunn*

CD91 The Role of Friendship in Psychological Adjustment, *Douglas W. Nangle, Cynthia A. Erdley*

CD90 Symbolic and Social Constraints on the Development of Children's Artistic Style, *Chris J. Boyatzis, Malcolm W. Watson*

CD89 Rights and Wrongs: How Children and Young Adults Evaluate the World, *Marta Laupa*

CD88 Recent Advances in the Measurement of Acceptance and Rejection in the Peer System, *Antonius H. N. Cillessen, William M. Bukowski*

CD87 Variability in the Social Construction of the Child, *Sara Harkness, Catherine Raeff, Charles M. Super*

CD86 Conflict as a Context for Understanding Maternal Beliefs About Child Rearing and Children's Misbehavior, *Paul D. Hastings, Caroline C. Piotrowski*

CD85 Homeless and Working Youth Around the World: Exploring Developmental Issues, *Marcela Raffaelli, Reed W. Larson*

CD84 The Role of Peer Groups in Adolescent Social Identity: Exploring the Importance of Stability and Change, *Jeffrey A. McLellan, Mary Jo V. Pugh*

CD83 Development and Cultural Change: Reciprocal Processes, *Elliot Turiel*

CD82 Temporal Rhythms in Adolescence: Clocks, Calendars, and the Coordination of Daily Life, *Ann C. Crouter, Reed W. Larson*

CD81 Socioemotional Development Across Cultures, *Dinesh Sharma, Kurt W. Fischer*

CD80 Sociometry Then and Now: Building on Six Decades of Measuring Children's Experiences with the Peer Group, *William M. Bukowski, Antonius H. Cillessen*

CD79 The Nature and Functions of Gesture in Children's Communication, *Jana M. Iverson, Susan Goldin-Meadow*

CD78 Romantic Relationships in Adolescence: Developmental Perspectives, *Shmuel Shulman, W. Andrew Collins*

CD77 The Communication of Emotion: Current Research from Diverse Perspectives, *Karen Caplovitz Barrett*

CD76 Culture as a Context for Moral Development, *Herbert D. Saltzstein*

CD75 The Emergence of Core Domains of Thought: Children's Reasoning About Physical, Psychological, and Biological Phenomena, *Henry M. Wellman, Kayoko Inagaki*

CD74 Understanding How Family-Level Dynamics Affect Children's Development: Studies of Two-Parent Families, *James P. McHale, Philip A. Cowan*

CD73 Children's Autonomy, Social Competence, and Interactions with Adults and Other Children: Exploring Connections and Consequences, *Melanie Killen*

CD72 Creativity from Childhood Through Adulthood: The Developmental Issues, *Mark A. Runco*

CD69 Exploring Young Children's Concepts of Self and Other Through Conversation, *Linda L. Sperry, Patricia A. Smiley*

CD67 Cultural Practices as Contexts for Development, *Jacqueline J. Goodnow, Peggy J. Miller, Frank Kessel*

CD65 Childhood Gender Segregation: Causes and Consequences, *Campbell Leaper*

CD46 Economic Stress: Effects on Family Life and Child Development, *Vonnie C. McLoyd, Constance Flanagan*

CD40 Parental Behavior in Diverse Societies, *Robert A. LeVine, Patrice M. Miller, Mary Maxwell West*

United States Postal Service

Statement of Ownership, Management, and Circulation

1. Publication Title: New Directions For Child & Adolescent Development
2. Publication Number: 1 5 2 0 - 3 2 4 7
3. Filing Date: 9/26/02
4. Issue Frequency: Quarterly
5. Number of Issues Published Annually: 4
6. Annual Subscription Price: $79.00 Individual, $180.00 Institution
7. Complete Mailing Address of Known Office of Publication (Not printer) (Street, city, county, state, and ZIP+4)
989 Market Street
San Francisco, CA 94103-1741
San Francisco County

Contact Person: Joe Schuman
Telephone: 415 782 3232

8. Complete Mailing Address of Headquarters or General Business Office of Publisher (Not printer)
Same as above

9. Full Names and Complete Mailing Addresses of Publisher, Editor, and Managing Editor (Do not leave blank)

Publisher (Name and complete mailing address)
Jossey-Bass, A Wiley Company
Above Address

Editor (Name and complete mailing address)
William Damon
Stanford Center On Adolescence
Cypress Bldg. C - Stanford University
Stanford, CA 94305-4145

Managing Editor (Name and complete mailing address)
None

10. Owner (Do not leave blank. If the publication is owned by a corporation, give the name and address of the corporation immediately followed by the names and addresses of all stockholders owning or holding 1 percent or more of the total amount of stock. If not owned by a corporation, give the names and addresses of the individual owners. If owned by a partnership or other unincorporated firm, give its name and address as well as those of each individual owner. If the publication is published by a nonprofit organization, give its name and address.)

Full Name	Complete Mailing Address
John Wiley & Sons Inc.	111 River Street Hoboken, NJ 07030

11. Known Bondholders, Mortgagees, and Other Security Holders Owning or Holding 1 Percent or More of Total Amount of Bonds, Mortgages, or Other Securities. If none, check box → ☐ None

Full Name	Complete Mailing Address
Same as Above	Same As Above

12. Tax Status (For completion by nonprofit organizations authorized to mail at nonprofit rates) (Check one)
The purpose, function, and nonprofit status of this organization and the exempt status for federal income tax purposes:
☐ Has Not Changed During Preceding 12 Months
☐ Has Changed During Preceding 12 Months (Publisher must submit explanation of change with this statement)

PS Form 3526, October 1999 (See Instructions on Reverse)

13. Publication Title: New Directions For Child And Adolescent Development

14. Issue Date for Circulation Data Below: Summer 2002

15. Extent and Nature of Circulation		Average No. Copies Each Issue During Preceding 12 Months	No. Copies of Single Issue Published Nearest to Filing Date
a. Total Number of Copies (Net press run)		1,097	1,010
b. Paid and/or Requested Circulation	(1) Paid/Requested Outside-County Mail Subscriptions Stated on Form 3541. (Include advertiser's proof and exchange copies)	354	341
	(2) Paid In-County Subscriptions Stated on Form 3541 (Include advertiser's proof and exchange copies)	0	0
	(3) Sales Through Dealers and Carriers, Street Vendors, Counter Sales, and Other Non-USPS Paid Distribution	0	0
	(4) Other Classes Mailed Through the USPS	0	0
c. Total Paid and/or Requested Circulation [Sum of 15b. (1), (2),(3),and (4)]		354	341
d. Free Distribution by Mail (Samples, complimentary, and other free)	(1) Outside-County as Stated on Form 3541	0	0
	(2) In-County as Stated on Form 3541	0	0
	(3) Other Classes Mailed Through the USPS	1	1
e. Free Distribution Outside the Mail (Carriers or other means)		75	75
f. Total Free Distribution (Sum of 15d. and 15e.)		76	76
g. Total Distribution (Sum of 15c. and 15f)		430	417
h. Copies not Distributed		667	593
i. Total (Sum of 15g. and h.)		1,097	1,010
j. Percent Paid and/or Requested Circulation (15c. divided by 15g. times 100)		82%	82%

16. Publication of Statement of Ownership
☐ Publication required. Will be printed in the Winter 2002 issue of this publication. ☐ Publication not required.

17. Signature and Title of Editor, Publisher, Business Manager, or Owner
Susan E. Lewis
VP & Publisher - Periodicals
Date: 9/26/02

I certify that all information furnished on this form is true and complete. I understand that anyone who furnishes false or misleading information on this form or who omits material or information requested on the form may be subject to criminal sanctions (including fines and imprisonment) and/or civil sanctions (including civil penalties).

Instructions to Publishers

1. Complete and file one copy of this form with your postmaster annually on or before October 1. Keep a copy of the completed form for your records.
2. In cases where the stockholder or security holder is a trustee, include in items 10 and 11 the name of the person or corporation for whom the trustee is acting. Also include the names and addresses of individuals who are stockholders who own or hold 1 percent or more of the total amount of bonds, mortgages, or other securities of the publishing corporation. In item 11, if none, check the box. Use blank sheets if more space is required.
3. Be sure to furnish all circulation information called for in item 15. Free circulation must be shown in items 15d, e, and f.
4. Item 15h., Copies not Distributed, must include (1) newsstand copies originally stated on Form 3541, and returned to the publisher, (2) estimated returns from news agents, and (3), copies for office use, leftovers, spoiled, and all other copies not distributed.
5. If the publication had Periodicals authorization as a general or requester publication, this Statement of Ownership, Management, and Circulation must be published; it must be printed in any issue in October or, if the publication is not published during October, the first issue printed after October.
6. In item 16, indicate the date of the issue in which this Statement of Ownership will be published.
7. Item 17 must be signed.

Failure to file or publish a statement of ownership may lead to suspension of Periodicals authorization.

PS Form 3526, October 1999 (Reverse)